World University Library

The World University Library is an international series
of books, each of which has been specially commissioned.
The authors are leading scientists and scholars from all over
the world who, in an age of increasing specialization, see the
need for a broad, up-to-date presentation of their subject.
The aim is to provide authoritative introductory books for
students which will be of interest also to the general
reader. Publication of the series takes place in Britain,
France, Germany, Holland, Italy, Spain, Sweden and
the United States.

Frontispiece Nicholas Kratzer, mathematician,
astronomer and a skilled maker of sundials,
was a friend of Erasmus, and of Holbein,
who painted this portrait.

Hugh Kearney

Science and Change
1500-1700

World University Library

McGraw-Hill Book Company
New York Toronto

For Martha, Jamie and Peter

© Hugh Kearney 1971
Library of Congress Catalog Card Number: 76-96433
All rights reserved. No part of this publication may be reproduced,
stored in a retrieval system, or transmitted, in any form or by any
means, electronic, mechanical, photocopying, recording or otherwise,
without the prior permission of the copyright owner.
Reprinted 1974
Photoset by BAS Printers Limited, Wallop, Hampshire, England
Manufactured by LIBREX, Italy

Contents

Introduction

The Scientific Revolution of the sixteenth and seventeenth centuries is now generally recognised as a decisive turning point in world history. It has taken its place in the judgment of most historians beside such movements as the Renaissance and Reformation, from which indeed it cannot be entirely dissociated. The innovations which it introduced are seen as a major cause of the transition from traditional modes of thinking, in which authority was accepted as natural and desirable, to 'modernity', in which critical assessment of all assumptions is encouraged as an essential part of maturity. The Scientific Revolution initiated scrutiny of nature and human nature by methods of hypothesis and experiment which were expected to lead to novelty and change.

This is of course an over-simplification. There had been 'Scientific Revolutions' before in the history of the world. By any reckoning, the Neolithic Revolution of 4000 BC represented a qualitative change in man's approach to his natural environment. Between the third and thirteenth centuries AD the Chinese made extraordinary headway in their empirical understanding of the universe. But it was the Greeks of *c*. 500-200 BC, more than any other group, who pressed beyond the accepted frontiers of knowledge to reach revolutionary interpretations of nature.

Chronologically speaking, the Scientific Revolution of Classical and Hellenistic Greece does not fall within the scope of this book, but because of its later influence in Europe it cannot be ignored. The mathematical achievement of Pythagoras (582-500 BC), the speculations of Plato (427-347 BC), the empiricism of Aristotle (384-322 BC), the geometry of Euclid (*c*. 300 BC), the engineering insights of Archimedes (287-212 BC), the astronomical observations of Ptolemy (*fl*. AD 139-161), the anatomical and medical work of Galen (*c*. AD 130-201) – all this was to be rediscovered in the West from the twelfth century onwards, after being lost sight of during the 'Dark Age' which followed the fall of the Roman Empire in the fifth century.

The rediscovery of Greek science was a complex process which stretched over five centuries, from the twelfth to the sixteenth. It began with the revival of Aristotelian logic in the twelfth century and the incorporation of further sections of Aristotelian science into Christian philosophy. Of those who took part in the work of 'baptising' Aristotle, the best known was the thirteenth-century theologian Thomas Aquinas (1226-74), but there were many others. These attempts to reconcile Aristotelian science with Christian doctrine were given the general name of 'scholasticism' by later generations of philosophers.

Theologians concentrated their attention upon Aristotle's logic and general philosophy. For others, however, Aristotle's empirical observations, together with Galen's medical works, were more important. Another focus of interest was in the fields of astronomy and astrology, which then were related to medicine because the planets were thought to influence the course of human life; hence, horoscopes provided crucial information for physicians or surgeons (e.g. whether it was safe to operate on a particular day).

By 1500, the assimilation of Aristotle, Galen and Ptolemy was complete and their views had been largely incorporated along with Christian doctrine in a vast synthesis, backed by the resources of Church and State. God, man, angels, animals, planets, and elements, all had their place in a world where man and the earth were at the centre, and heaven beyond its circumference. This view of the world was emotionally satisfying, religiously orthodox and poetically inspiring, but it was to be overthrown in a remarkably short time. Within less than two centuries almost every assumption which had been accepted since 500 BC and which the west had painfully relearned since the twelfth century was questioned.

The most striking change occurred in cosmology. From the mid-sixteenth century it was argued that the earth no longer occupied its traditional position at the centre of the universe but was merely one of several planets revolving round the sun. Certain Greek scientists, notably Aristarchus (his heliocentric theory was known from a single sentence), had held this view, but its exposition in modern times was due to mathematicians of genius, Copernicus, Galileo, Kepler, Descartes and Newton. Their success led to an ever-increasing acceptance of mathematical analogies in fields outside astronomy.

Another decisive change was taken when western European scientists began to make observations on their own account, to construct novel hypotheses and to devise new experiments. This began piecemeal in the fourteenth century, but it made real headway in the sixteenth. By the end of that century, the number of modern scientific contributions increased still further, and more dramatically again, during the seventeenth century. The novel type of experimental method also originated mainly with mathematicians, including Galileo, Pascal, and Newton.

The Scientific Revolution, though it originated in the recovery of Greek science, led to the overthrow of the Greek way of interpreting the universe. By 1700, Descartes and Newton had replaced Ptolemy in astronomy, Galileo replaced Aristotle in physics, Vesalius and Harvey replaced Galen in medicine. In mathematics, the moderns made notable advances in new fields, especially in algebra and co-ordinate geometry. Logarithms were invented at this time (*c.* 1610-20).

But the scientific achievement which most struck contemporaries was that of Isaac Newton, whose *Principia* brought together in a mathematical synthesis the courses of the planets and the path of a falling stone. Newton destroyed completely one of the basic assumptions made by the Greeks: that the celestial and terrestrial worlds were different in nature and hence in the natural laws which applied to them. Newton in his *Optics* showed that white light was composed of coloured rays, and demonstrated dramatically the value of experimental method in reaching revolutionary conclusions.

Looked at in the context of world history, the Scientific Revolution was an extraordinary intellectual leap which had repercussions ultimately on every aspect of western thought and life. A new tradition had been created which was to bear fruit a thousandfold in the eighteenth and nineteenth centuries, though by 1900 the assumptions of the original Scientific Revoltuion had been modified almost out of recognition.

This chronological survey of the Scientific Revolution will serve our purposes as a simple account of what happened in the history of science between 500 BC and AD 1500. But it is open to the criticism of being too simple. It brings out the point that the Scientific Revolution was at once a recovery of Greek thought and a repudiation of it. What it does not emphasise is the complexity of the whole process and its links with religion and philosophy.

Greek thought, after all, was not a unified system, any more than western thought is today. Its various schools of thought held sharply conflicting opinions, which changed over a thousand years of development. After AD 500 western Europe lapsed into semi-barbarism (a state of society not without its compensations in art) and intellectual revival began seriously only *c*. AD 1000. Thus the task of recovering Greek thought was not a relatively simple problem, akin to a man putting together the displaced pages of an encyclopedia. It was the painful effort of a largely barbarian society coming to terms with the intellectual sophistication of a superior civilisation. This was in any event a task of immense difficulty. It was intensified by the fact that Greek thought became available from the twelfth century onwards only in defective Latin translations.

Another problem was the reconciliation of Greek thought with Christian tradition. In broad terms this meant bringing together in a single whole, views based upon the amalgam of Jewish history and poetry called the Bible (in the Greek translation known as the Septuagint, which dated from *c*. 200-100 BC) with the philosophy and science of the advanced urban civilisation of Greece – a formidable task. Medieval scholars had one great advantage in all this: they had no sense of history. They did not distinguish chronologically between the Greek thought of one period and that of another, still less between Greek and Hebraic modes of thought. Instead, they attempted to create a logical construction of views which were historically distinct. Often they accepted as genuine what later were seen to be forgeries.

The whole range of Greek thought did not become available to the West immediately. During the Middle Ages the science and philosophy of Aristotle was studied, beginning with his logic and followed by his physics, metaphysics and biological treatises. The work of Plato was almost unknown. Had Plato's critical open-ended dialogues been known first, it is arguable that western science would have taken a different course, more mathematical and less empirical. But it was only in the late fifteenth century that Platonism once more became

Signs of the zodiac from an illuminated manuscript. Gemini ('the twins') and Sagittarius ('the archer') are shown here. The forecasting of horoscopes was very much part of an astronomer's activities even for Tycho Brahe and Kepler.

a force to be reckoned with, and even then only in the form of neo-Platonism.

Plato himself wrote in the fourth century BC within the social world of the Greek City State. Neo-Platonism, with its vision of the world as a series of emanations from the Divine Mind, was the product of the later centuries of the Roman Empire (AD 200-400) and was mystical and anti-rational in tone. It was welcomed with great enthusiasm during the Renaissance as a more religious philosophy than Aristotelianism partly because it presented a view of the world in which miracles did not seem out of place.

The recovery of most of Greek thought was completed during the Renaissance period, when the atomism of Democritus (fl. 470 BC) became known as well as the technical treatises of Archimedes and others. Taken together these provided the basis of alternative world-views to both Plato and Aristotle. In particular, they made available the concept of a mechanistic theory of the universe.

The conflicting heritage of Greek thought created a challenge for western thinkers. It also led to immense intellectual confusion, which was intensified by the Reformation and the struggles of sixteenth-century Europe when religious orthodoxies struggled for dominance. From this the history of science was not immune. Science did not develop in a separate compartment labelled 'The Scientific Revolution', but was itself part of the process of social and intellectual change. The rise of mathematics and the development of experimental method took place in a world where religion and science (or 'natural philosophy' as it was termed) were not distinct activities, as they are in the West today. Science arose in conditions of confusion, suspicion and irrationalism, as did other activities of the period. Thus, upon our first simple view

of a man

weye

ho so

and se

wo and

passed

in yo

ry an

the w

it was

closyd

for as mko as Evns

of the Scientific Revolution, we must impose another which takes due account of this complexity. The scientists whose work will be discussed, must be seen as working within an intellectual environment which was dominated by three varying patterns of Greek thought, seen through the prism of two thousand years.

1 Three traditions in science

In this book, I shall argue that the key to interpreting the origins and course of the Scientific Revolution is to be found in three distinctive traditions or paradigms – the organic, magical and mechanistic. Before I discuss this in detail, however, I would first like to offer some criticisms of a historical outlook which tends to dominate and distort general accounts of the Scientific Revolution and which I will call 'the Whig interpretation of history'.

The Whig interpretation of history

This interpretation implies a view of the past which divides men essentially into two simple categories, progressive or reactionaries, forward-looking or backward-looking, Protestants or Catholics. As Sir Herbert Butterfield pointed out in a brilliant essay, this way of looking at history leads to gross distortions because it imposes the standards of the present upon the past. One danger lies in the assumption that the purpose of the past has been to prepare the way for the present. Another is to trace a simple line of continuity from past to present. Perhaps the basic error is to substitute explanation based on logical progression for a less rational and more complex interpretation of the past.

The Whig interpretation takes its name from the account of English constitutional history given by nineteenth-century Whig historians, who saw English liberty 'slowly broadening down from precedent to precedent'. The foundations of freedom were laid with Magna Carta in 1215, rescued in the Civil War of the seventeenth century, and confirmed by the Glorious Revolution of 1688, the implications of which were worked out in the Reform Acts of the nineteenth century. The basic assumptions which lay behind this interpretation of English

The Copernican universe with the sun at the centre
from *De Revolutionibus Orbium Coelestium*, Nuremberg
(1543). Copernicus retained an immobile sphere of stars
beyond the planets, of which the earth was now one.
A comparison between this and the illustration on
page 31 epitomises the Copernican revolution.

history was a simple one. Englishmen were divided into two
simple categories, those who loved liberty, the Whigs, and
those who did not, the Tories. By this test, English history
assumed an intelligible pattern, which ultimately was enshrined
as a myth.

But it would be misleading to restrict our attention to
England and the Whig historians. Most nationalist interpretations of history make similar judgments about the past.
History appears as a story of those who supported the rise
of the nation and those who did not. Crucial differences between the patriots of one generation and another are lost sight
of and essential distinctions between men of the same generation are slurred over. Nationalist history of this kind seems
to be the response of an overpowering emotional need within
newly established political societies, which feel the need
to create a 'past' for themselves.

Perhaps the most influential example of the Whig interpretation of history today is provided by Marxists. This fact
is glossed over by the Marxist claim to be writing 'scientific'
history and by the undoubted erudition and imagination of
many Marxist historians. But behind much Marxist history
lies the assumption that history is a record of progress, with
progressives on one side and reactionaries on the other. Great
play is made with such terms as 'forward-looking' and 'backward-looking' as if it were possible to apply such concepts to
the complexities of the past, as if indeed these concepts had
meaning of an ascertainable kind. We must group such
Marxists with the Whigs, as historians who are committed
to seeing the past in terms of the present, and to applying what
are in effect two simple moral categories to history.

What has all this to do with the history of science? I believe,
everything. For the most part, general histories of science have

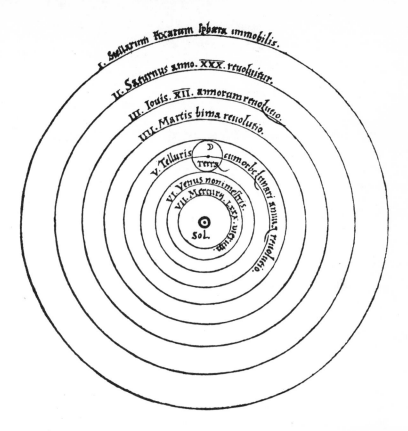

been written on the assumption that beneath all the technical detail there was a simple tale to be told. From this point of view, the history of science appears as the story of the emergence from clouds, irrationalism or superstition of a rational method of interpreting nature. The early scientists are seen as groping for the discovery of a scientific method which was to be finally revealed in all its clarity by later generations. If these early scientists were opposed in their own day, the reason was simple: they were the progressives and their opponents were the reactionaries.

Reduced to skeleton form, the Whig interpretation of the history of science runs as follows. The first substantial break-

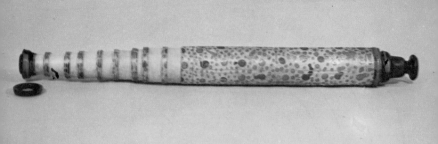

through was made by Copernicus, the Polish astronomer (1473-1543), who put forward the theory that the earth went round the sun instead of the sun going round the earth, as the conservatives thought. His view was taken up at the end of the century by the German scientist Johannes Kepler (1571-1630) and by the Italian Galileo Galilei (1564-1642). Kepler discovered that the paths of the planets were elliptical not circular, and Galileo was the first to use the telescope in astronomy and to formulate the mathematical law of falling bodies. Finally came Sir Isaac Newton (1642-1727) who in his *Principia* brilliantly applied the same law to planetary motion and to falling bodies alike.

This essentially is the Whig version of the Scientific Revolution, the basic structure underlying for example such books as E.J.Dijksterhuis's monumental study *The Mechanisation of the World Picture* (1950). Copernicus, Galileo, Kepler and Newton are the key names, but other scientists may be fitted into the scheme of things – Tycho Brahe, Descartes, Robert Boyle and Leibniz – once the initial judgment has been made that they are 'forward-looking'. Those who do not fit into these categories are consigned to oblivion. (Even Butterfield on occasion cannot avoid using harsh judgments like 'lunacy' about the views of men who do not fit into his interpretation.)

The attractions of this simple model are obvious. It reduces a very technical piece of intellectual history to manageable proportions. As a logical structure, stressing the logical progression of ideas, it makes an appeal to the philosophical and scientific mind. But it is in fact open to exactly the same objections as the Whig interpretation proper, or to nationalist history, or to Marxist history. The proof of the pudding is in the eating. A Whiggish history of science breaks down like other Whig interpretations as soon as detailed research into specific problems or particular periods is undertaken.

Indeed the Whig interpretation of the history of science no longer seems convincing precisely because of the amount of detailed historical research in recent years which has appeared in such periodicals as *Isis*, *Ambix*, *Archives Internationales pour l'histoire des Sciences* and *Journal of the History of Ideas*. It is perhaps true to say that a greater amount of distinguished work has appeared in print since the Second World War than ever before. But even apart from this, there is a general dissatisfaction with the artificial confines of Whig interpretations. Hence, many historians following the example of the French school of Marc Bloch and Lucien Febvre have moved towards 'histoire intégrale'. History written on these lines aims at total history, in order to make intelligible all aspects of life in particular societies, not merely a single line of apparent progress.

To some extent, the Marxists had already pointed the way in this direction. Indeed, it may be said that one of the beneficial effects of Marxist interpretations of history has been to force historians to take the broad view, and to see interconnexions between apparently disparate topics. Unfortunately, however, Marxism in practice is often too rigid and too predictable, providing only one form of explanation and one kind of analogy. A more fruitful example has been set by sociologists and social anthropologists. We are now less tempted to make simple distinctions between rational and irrational behaviour. We look instead for the social function of certain modes of activity. We see that what is called 'magic' in a primitive society may correspond to what is called 'science' in a more sophisticated one. We are also introduced to a wider range of concepts than is supplied in a Marxist 'class' analysis, such as 'roles', 'function', 'status' and other concepts. And what applies to the political and intellectual historian applies with equal force to the historian of science. Thus the history of science no longer appears as a self-contained activity, an enclosed scientific tradition, with truth 'slowly broadening down from precedent to precedent'. Historians of science are now engaged in seeking the influence of allegedly non-scientific and non-rational factors upon scientists.

To some extent, this approach has been practised for a long time. Darwinism is an obvious case in point. It seems certain that Darwin's scientific imagination was kindled by the Malthusian law of population. Malthus, in stressing the continual pressure of population upon food supply, provided the key for Darwin's theory of natural selection. In other words, to explain the origins of *The Origin of Species*, the historian of science must come to terms with the general history of the early nineteenth century.

The three scientific traditions

In attempting to escape from the bonds of a Whig interpretation of the Scientific Revolution I will argue that there were, during this period, at least three approaches to nature which may be broadly termed 'scientific' in the sense that they all produced discoveries which have been incorporated within the modern scientific tradition. But 'modernity' is a dangerous criterion. All three of them were bound up with religious assumptions about the universe, whereas modern science by definition is a secular activity. No exponent of a particular tradition had a concept of science in its modern sense, indeed, the term 'scientist' was first invented in the nineteenth century. Hence, the exploration of natural phenomena during our period must be seen according to its own terms of reference, even though much of it may seem 'magical' or 'superstitious' to us. There is no direct line of progression from them to us, and whenever we are tempted to identify our own concerns with one or other of these traditions, we are victims of an optical illusion which further analysis will expose.

In general terms the three traditions may be described as organic, magical and mechanistic. Within the organic tradition the scientist explained the natural world in terms of analogies drawn from what we now call biology. The language which he used originated in observation of growth and decay, with the analogy of the acorn growing into the oak always ready to hand. Thus the veins of metallic ore were accounted for by the explanation that the metal had 'grown' in a favourable place. What struck this type of mind about nature was not its regularity and uniformity, but its constant change. Yet within the process of change there was a consistency which had to be accounted for. Acorns did not grow into chickens. This led to the view that there was a potentiality or purpose built into all natural phenomena, a so-called 'final cause', which dominated development.

Within the organic tradition, the scientist turned almost inevitably to the study of living organisms. And even when he dealt with what we would now regard as inanimate nature, he tended to attribute life to it or to use language and terms

The salamander from the *History of Animals* by the Swiss naturalist Conrad von Gesner (1516-65). The title shows that it was intended to emulate Aristotle's work of the same name. The salamander was believed by many to live in fire because of the presumed coldness of its temperament. But J. P. Wurffbain in his *Salamandrologia* (1683) disposed of this myth.

derived from his primary interest in life and growth. The terms 'natural' and 'unnatural' were applied within the organic tradition to problems of motion. A falling stone was behaving 'naturally', a projectile hurled upwards was moving 'unnaturally'.

A second tradition, the magical, provided a scientific framework in which the world of nature was seen as a work of art. (I use the word 'magical' in preference to 'aesthetic' because it suggests the overtones of mystery which I think were involved.) The appropriate analogies and the language of the scientist derived from a view of nature in which beauty, contrivance, surprise and mystery were seen as its dominant characteristics. Within this general framework, however, there was room for an immense variety of emphasis. Some interpreters turned to mathematics, and to a world which was presumed to lie beyond the constant change of the observable world. Others saw the role of the interpreter of nature as akin to that of the magician, whose possession of the secrets of nature brought him power.

Within the magical tradition, the Christian deity took on some of the attributes thought appropriate to a magician or artist, and the scientists who worked on these lines saw themselves as following the creator's example and, by the pursuit of clues in the world of nature, gaining an insight into the 'divine artist's' mind.

The third tradition rested upon a view of nature in which the dominant analogy was the machine. What struck scientists who worked within this framework was the regularity, permanence and predictability of the universe. The planets were seen in mechanical terms, as were also the human body, the animal kingdom and even the processes of artistic creation. From this point of view, the Christian God took on some of

Figura prior ad viuum expreſſa eſt. altera vero quæ ſtellas in dorſo gerit, in libris quibuſdam publicatis repe-
ritur, conficta ab aliquo, qui ſalamandram & ſtellionem à ſtellis dictum, animal vnum putabat, vt coniicio. &
cùm à ſtellis ſtellionem dictum legiſſet, dorſum eius ſtellis inſignire voluit.

the characteristics of an engineer. Mechanists concentrated upon those aspects of the world which were most easily explained in mechanical terms. Questions which were thought marginal within the organic and aesthetic traditions, such as problems of acceleration, took on a new significance within the mechanistic framework. The concept of unchanging scientific laws, expressible in mathematical terms, was of particular importance in this tradition and a mathematical approach came to be its dominant characteristic.

In historical terms we may think of the magical tradition as being a reaction against the organic tradition, and the mechanical as a reaction against magic. But it must be said that within each tradition there were sub-groups and distinctive schools of thought. What we have done in effect is to construct three models or paradigms which explain many aspects of the course of the Scientific Revolution, but, as we will see, each tradition was related to some aspect of Greek thought, the organic tradition to Aristotle, the magical tradition to neo-Platonism and the mechanist tradition to the atomists and Archimedes.

The organic tradition

The organic tradition in science rested upon a threefold base of Aristotle, Galen and Ptolemy – and of these the greatest was Aristotle. Aristotle's biological treatises, Galen's medical observations and Ptolemy's great astronomical corpus, the *Almagest*, provided a mass of empirical data which was unrivalled over a thousand years after it had been produced. The sheer bulk of this work gave confidence to scientists within the organic tradition and made it possible for them to dismiss objections as marginal. If we look at the Aristotelians through Galileo's eyes we see a group of simple-minded theorisers. In their own estimation, which was not without justification, they were the empiricists.

Aristotle's own empirical treatises, notably the *Natural History of Animals*, showed powers of patient observation combined with a healthy distrust of excessive theorising. To the influence of this aspect of Aristotelianism was added the empiricism which Galen displayed in his anatomical treatises. Ptolemaic astronomy may also be regarded as the record of observations made in a celestial laboratory, where the 'experiments' were repeated endlessly under the same conditions.

But the organic tradition was something more than a collection of scientific observations. It was also a philosophical system, extending into metaphysics, ethics and logic, which within most European universities during most of our period (1500-1650) was thought to provide the only acceptable synthesis of human knowledge, even though it might be open to modification in detail. Thus the organic tradition served two inter-connected purposes; it was a source of scientific information and it provided a pattern of intellectual coherence.

Aristotelian science, as expounded in countless text books

of the sixteenth and seventeenth centuries, stressed the role of purposive development in the world. Change was a constant feature in nature but it was change controlled by the end in view (or final cause). In this emphasis we may see the impact of Aristotle's own biological researches, which he used as the key to other sciences. In Aristotelian science the dominant analogy was provided by natural growth, which, in Aristotelian terms, was movement directed towards an end. Aristotelians saw this process repeated throughout nature, not merely in living things but in the movement of inanimate objects and in 'chemical' change.

This was not entirely an academic point. Aristotle looked upon his scientific approach as a conclusive answer to the mechanistic assertions of Democritus. Galen, five centuries later than Aristotle, had also attacked the mechanism of his contemporaries. Thus from the first the organic tradition was a series of entrenched theoretical positions, which were anti-mechanist in spirit. We can understand the renewed appeal which this made in the sixteenth century when Greek mechanistic doctrines enjoyed a new lease of life and threatened the basis of Christian belief in providence.

Aristotelian theories of physics and of chemical change were bound up with Aristotelian cosmology. The earth was at the centre of the universe and around it revolved the planets and the sun, each in their sphere. There was an absolute up and down and there was also a complete division between the lunar world and the sub-lunar world, each of which had its own physics and its own 'chemistry'. In the lunar world, the planets moved in circular orbits and their composition was of an incorruptible element. In the sub-lunar world, change was a constant feature, motion was rectilinear and matter was composed of four elements.

Two of the three liberal arts, Grammar and Logic, which, with Rhetoric, formed the trivium. In the medieval world this was regarded as the educational foundation leading eventually to theology. The tower in the illustration below shows the gradation from the liberal arts to divine truth. The role of logic is suggested in the illustration on the right, 'Typus Logice'. Armed with an axe and a sword, symbolising the techniques of question and

syllogism, the student makes for a wood of opinion. He steps
gingerly through a morass of fallacies, following the two hounds
truth and falsehood. In the wood, various schools of thought are
also indicated (e.g. Occamists and Thomists). These illustrations
from an early sixteenth-century book *Margarita Philosophica*
show how influential scholasticism was *c.* 1500–1600.

The Pre-Copernican universe, showing the earth at the centre, with Aristotle's four elements, Earth, Air, Fire and Water, surrounded by the spheres of the planets and the sphere of the fixed stars. From *Practica compendiosa artis Raymond Lull* (1523).

It is easy to dismiss this picture as fanciful, but we must try to understand the reasoning which lay behind it. The astronomical observations of Ptolemy's *Almagest* supported the view that the planets moved at varying speeds in circular orbits, with the earth at the centre of the cosmos. Ptolemy in 1500 could be quoted in support of Aristotelianism, not in poetic terms but as the source of the most advanced astronomical observations available. Aristotelians used this technical information as support for their philosophical principles by which the circular motion of the planets, eternal and unchanging, was perfect motion, to be contrasted with the limited rectilinear and hence imperfect motion of bodies on earth.

Within the imperfect sub-lunar world, different scientific principles operated. From Empedocles onwards (*c.* 450 BC) the world of matter was assumed to be composed of four elements, earth, air, fire and water, two of which were 'heavy' (earth and water) and two 'light' (fire and air). On this basis, the Aristotelians explained both 'physics' and 'chemistry'. Motion was the movement of heavy or light elements to their natural place in the universe. Objects in which earth or water were predominant moved downwards, whereas the contrary occurred with fire and air (e.g. smoke, clouds etc.). This conclusion rested upon rudimentary observation. It also fitted in with the Aristotelian view that movement towards an end was the dominant feature of the universe. A falling body fell at increasing speed because it was 'seeking' its natural position. It fell in a straight line because this was the form of limited motion appropriate to an imperfect world.

Aristotelians did not exclude mathematical considerations from their general explanation of falling bodies. Following Aristotle himself, most of them took the view that the speed of a falling body was proportional to its weight, hence lead

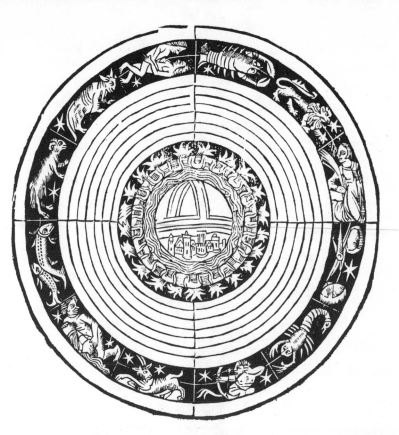

fell many times faster than a feather. Experiments in water 'proved' the point with some justification, experiments in air were more ambiguous. From the fourteenth century, some Aristotelians, such as the Frenchman Oresme (+ 1382) tried to discuss the behaviour of falling bodies in geometrical terms.

The 'difficult case' for Aristotelian theory was the behaviour of a projectile which followed a course of 'unnatural' motion against its 'natural' proclivity to return to earth. This was explained by most Aristotelians as the effect of the movement of the air. From the fourteenth century more sophisticated analysts explained it in terms of a quality ('impetus') which the projectile acquired and gradually lost.

An armillary sphere was a skeleton celestial globe. The one shown here illustrates the Copernican planetary system. It was made for Robert Boyle's nephew, Charles Boyle (1676-1731) fourth Earl of Orrery. Hence the name 'Orrery' often given to such spheres. By 1700 in England Copernicanism was universally accepted.

The four elements, earth, air, fire and water, provided the key within the organic tradition to the composition of matter. Here, as in physics, Aristotle's emphasis was anti-mechanist. In defending the existence of four elements, Aristotle refuted those who believed in a single element composed of mechanically interacting atoms. Aristotelians explained chemical change in terms of a changing composition of the four elements in a substance. (Thus charcoal on this view had lost much of its 'air' and hence had a higher 'earth' content than wood). But material change was not the only factor. It was accompanied by a change of 'substantial form', i.e. the qualities ('form') of wood were different from those of charcoal. Thus even in chemical change where only inanimate objects were concerned there was 'purposive' development in the transition from one form to another. Aristotelian emphasis on substantial form and qualitative difference ruled out the possibility of a mechanical explanation of chemical change. Even in physics 'impetus' was regarded as a quality possessed by a projectile.

Although we must admit that there was a hard core of empirical data at the centre of Aristotelianism, the fact remains that Aristotle and his followers could not resist systematising and generalising on a slender basis. The accepted instrument for such reasoning was the syllogism, which in its simplest form appeared as: 'All men are mortal, Socrates is a man, Socrates is mortal.' Using this, and elaborations of it, Aristotle constructed a philosophical system which is either imaginative or fanciful according to one's point of view. It led him to elaborate on the destruction between circular and rectilinear motion in the following way:

It can now be shown plainly that rotation is the primary locomotion. Every locomotion, as we said before, is either rotatory or rectilinear or a compound of the two: and the two former must be prior to the last, since they are the elements of which the latter consists. Moreover, rotatory locomotion is prior to rectilinear locomotion because it is more simple and complete, which may be shown as follows. The straight line traversed in rectilinear motion cannot be infinite: for there is no such thing as an infinite straight line; and even if there were, it would not be traversed by anything in motion: for the impossible does not happen and it is impossible to traverse an infinite distance. On the other hand, rectilinear motion on a finite straight line is, if it turns back a composite motion, in fact two motions, while if it does not turn back it is incomplete and perishable: and in the order of nature, of definition, and of time alike the complete is prior to the incomplete and the imperishable to the perishable. Again, a motion that admits of being eternal is prior to one that does not. Now rotatory motion can be eternal: but no other motion, whether locomotion or motion of any other kind, can be so, since in all of them rest must occur, and with the occurrence of rest the motion has perished. Moreover, the result at which we have arrived, that rotatory motion is single and continuous, and rectilinear motion is not, is a reasonable one. In rectilinear motion we have a definite starting-point, finishing-point, and middle-point . . . On the other hand, in circular motion there are no such definite points: for why should any one point on the line be a limit rather than any other? Any one point as much as any other is alike starting-point, middle-point, and finishing-point.[1]

The universal application of this type of reasoning, together with Aristotelian emphasis upon substantial forms and final causes, was the chief target for attack in the seventeenth century. Its critics complained that Aristotelianism in explaining everything, explained nothing. In defence it must be said that in 1600 there was no comparable system of scientific explanation. Aristotelianism provided a framework of discussion and a target to aim at. This paradigm with all its faults was better than no paradigm at all. Not until Descartes wrote his *Principia Philosophiae* (1644) was there an alternative.

Though Aristotle was not a Christian, and much of his teaching (e.g. the eternity of the world) was unacceptable to the Christian church, he had become by 1500 the dominant intellectual influence upon theology. Aristotelian terms such as substance and accident, or matter and form were used to explain Christian teaching on the Eucharist (e.g. transubstantiation), and the Aristotelian emphasis upon final causes helped to elucidate the operation of God in the world of nature. The god of the theologians, if not the Bible, was a deity whose mind was revealed in the purposive working of the universe. God was a logician whose premises could be scrutinised and his nature examined. The very working of divine grace was open to logical analysis. In this emphasis on purpose and logic, Aristotelian science and scholastic theology marched together.

Thus over and above all considerations of the intellectual and scientific calibre of Aristotle, Galen and Ptolemy, and the emotional satisfaction of explaining all natural phenomena, the strength of the tradition during our period lay in its close association with religious orthodoxy. In all Catholic and most Protestant universities during much of the sixteenth and seventeenth centuries, scholasticism gained ground. The organic tradition was in power, so to say, backed by the resources of church and state and strongly entrenched in the universities. The two rival traditions operated at the margin, under pressure if not persecution.

Despite the entrenched position which they enjoyed within the establishment, the Aristotelians could not suppress all criticism and they found themselves attacked on several crucial issues – the geocentric theory of the universe, the im-

This mosaic of Hermes Trismegistus, dating from the 1480s, is to be found in the cathedral at Siena. It bears witness to the strange vogue which Hermetic ideas enjoyed during the Italian Renaissance. Trismegistus, seen here with Moses (?) and an Egyptian personage, is described as the contemporary of Moses. For someone who never existed, he had an extraordinary hold on contemporary minds.

possibility of a vacuum and the course taken by a projectile. In all three instances, empirical observation showed Aristotelian theory to be wanting. If we ask why these issues came up rather than others, the answer is that Aristotle's deductive reasoning was particularly extended at these points and the divergence between theory and practice was most startling. If we ask why they came up at this time and not earlier, the key factor seems to be the availability of alternative scientific paradigms, themselves derived from Greek sources and thus as worthy of respect as Aristotle – the magical and mechanistic traditions. From within these new paradigms certain aspects of Aristotelianism seemed more open to criticism than others.

The magical tradition

The role of the scientist within this intellectual framework was to become attuned to the message of the universe and to be something of a magician, whereas the organic scientist was close to being a logician. God was a magician, a wonder-worker, not the rational first mover of Aristotle, and the best model for the scientists to follow was to become a mystic who could hear the magical music of the universe.

Much of the inspiration for this attitude came from the writings attributed to an ancient and mysterious Egyptian figure Hermes Trismegistus. Trismegistus (thrice blessed Hermes), who never in fact existed, was thought to be the author of over a dozen treatises (The Hermetic Writings) which claimed to expound the wisdom of the Egyptians during the period of Moses. The treatises first became available to the West after the fall of Constantinople (1453) and they were translated from the Greek by Marsilio Ficino (1433-95)

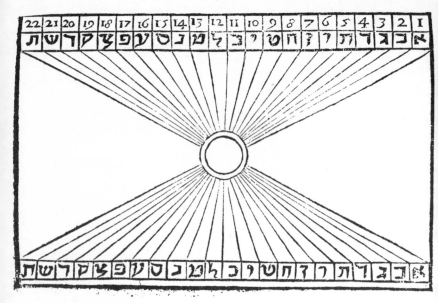

as a matter of urgency by order of Cosimo de Medici, who wished this task to be given precedence over the translation of Plato himself. From then onwards, until well into the seventeenth century, they exercised a powerful fascination over the western mind.

It is easy to see the reason for the influence of the Hermetic tradition. In Trismegistus, the Christian Church now had a source of wisdom which went back (or at least was believed to) beyond Plato to the original Mosaic revelation. Trismegistus was thought to have been the recipient of divine revelation about the physical world, as Moses had been about the moral world. From this point of view, the Egyptians were seen as the custodians of secular wisdom, as the Jews were of sacred wisdom. Hitherto, (it had been thought), the Greeks, in the persons of Pythagoras and Plato, had been the only source of access to Egyptian lore, and the West had only known it at second hand. Now, towards the end of the fifteenth century, the Egyptian treatises were available (or seemed to be) in their original form.

What was the message of the Hermetic writings? Among other things, they taught that the sun was the centre of the universe, round which the earth revolved. Light was the source of life. The sun was a symbol of the Godhead. The Hermetic treatises also incorporated certain Pythagorean assumptions, which stressed a mathematical harmony in the cosmos. The secrets of the cosmos had been written by God in a mathematical language, which could be discerned, for example, in musical harmonies.

The Hermetic writings provided the basis for a view of the cosmos which had implications for science and scientific method. It was a world full of magical powers, the secrets of which were open only to the chosen few who were willing to look beyond surface phenomena. The explorer of nature was an ascetic, studying the occult, within the confines of an esoteric community. The watchwords of this approach were mysticism, mystery and secrecy. There was thus a sharp contrast between the Hermetic and the Aristotelian 'scientist'.

Thanks to the work of the great classical scholar Isaac Casaubon (1559-1614), who first dated the Hermetic writings accurately, we now know that they originated in the second century AD. They belong in fact to the movement of mysticism and philosophy, known as neo-Platonism, founded by Plotinus (AD 205-270) and carried on by Porphyry (AD 232-303). Plato himself regarded the material world as 'unreal' (true reality lay in the unchanging world of forms) and Plotinus used this approach as the starting point for a philosophy in which the material world was the last and lowest form of being. Under the influence of Eastern mysticism, Plotinus believed that the source of being was 'the One' from which derived a trend of emanations, life, mind and soul, and finally, matter. For the neo-Platonist the human soul was spirit encased in matter,

Overleaf Two illustrations from Thomas More's *Utopia* (1515-16). It was in no sense a scientific treatise, but his life of Pico della Mirandola (1510) links him interestingly with the

whereas for the Aristotelian it provided 'form' to 'matter'. The neo-Platonic soul was imprisoned in the material world; the Aristotelian soul was an informing principle.

The late fifteenth century was marked as we have suggested earlier by a reaction against Aristotelian rationalism and its technicalities. The Hermetic Writings were only one among several treatises which were permeated with neo-Platonic influences. They included the Jewish Cabbala (literally – 'tradition'), which claimed to reveal the hidden secrets of the Old Testament by the use of ciphers. Among these 'secrets' was the neo-Platonic doctrine of the creation of the world by means of emanations from the Divine Being. In this atmosphere, the figure of Pythagoras took on a new significance, as the model of a mathematician who sought and found mystical combinations of numbers. Mathematics in this new view offered the key to a world of unchanging realities, close to, if not identical with, the Divine Mind. The pursuit of mathematics was not a secular activity. It was akin to religious contemplation. For Aristotelians, on the other hand, mathematics ranked low as an intellectual pursuit and had no religious connotation.

If neo-Platonism had remained the obsession of a few eccentric thinkers, there would be little point in discussing its significance in a book on the Scientific Revolution. In fact however the neo-Platonic approach made an enormous impact upon the intellectual world of the sixteenth century. It may be seen in More's *Utopia*, in the work of Pico della Mirandola and not least in the writings of Copernicus and Kepler. In the seventeenth century its influence extended to the Cambridge Platonists (more properly, the Cambridge neo-Platonists) and their greatest pupil, Sir Isaac Newton.

The neo-Platonic theory of matter offered an exciting

neo-Platonism of the Renaissance. The *Utopia* may be seen against this intellectual background, though this is not the only profitable perspective. The interest of the illustration on the left lies more in what it anticipates than in itself. In the seventeenth century, John Wilkins, among others, played with the notion of an artificial language which would be free from the traps of ordinary discourse as described by Bacon in the *New Organon*.

alternative to the prevailing Aristotelian orthodoxy of the four elements. Matter, for the neo-Platonist, was a link with the world of the spirit. Neo-Platonists held that the mineral and vegetable kingdoms offered reflections of spiritual realities. The 'microcosm' of this earth was believed to reflect the 'macrocosm' of greater reality. Here chemistry took on a quasi-religious aura which acted as an emotional stimulus towards novelty. Paracelsus for example, applied the neo-Platonic trend to chemical theory and the same approach may be seen in his seventeenth-century successor, Van Helmont (see page 126).

To men with imagination, the message of neo-Platonism offered a heaven-sent escape route from the rationalism of academic Aristotelianism. This was the sixteenth-century equivalent of Romanticism. Indeed we could do worse than look upon Shakespeare's *The Tempest* as an illustration of the appeal possessed by the Hermetic tradition. Prospero was the ideal type of the Hermetic scientist bringing justice and peace to a disturbed world, an approach which had great appeal in a century torn by religious bitterness.

The mechanist tradition

The magical tradition reached the peak of its influence at the end of the sixteenth century. From then on a reaction set in against it based upon a mechanistic view of the universe and propagated during the next century in the work of Mersenne, Hobbes and Descartes.

If we seek an origin for mechanism we may first consider the economic background of the period. There is a tempting connection to be made here between the increased use of machinery in the world of the sixteenth century and the mech-

VTOPIENSIVM ALPHABETVM.

a b c d e f g h i k l m n o p q r s t v x y

ÓΘΦΟϹΟϽϹᴖᴗƧᐃᒐᒪᒣꓶᴴⅢⅢⅠⅠ

Tetrastichon vernacula Vtopiensium lingua.

Vtopos   ha   Boccas   peu   la

chama   polta   chamaan

Bargol   he   maglomi   baccan

soma   gymno   sophaon

Agrama   gymnosophon   labarembacha

bodamilomin

Voluala   barchin   heman   la

lauoluola   dramme   pagloni.

Horum versuum ad verbum hæc est sententia.

Vtopus me dux ex non insula fecit insulam

Vna ego terrarum omnium absq; philosophia

Ciuitatem philosophicam expressi mortalibus

Libēter impartio mea, nó grauatim accipio meliora.

VTOPIAE INSVLAE FIGVRA

44

anical analogies to which Galileo and Mersenne turned. The American historian John U. Nef (1940), argued that there was an 'industrial revolution' in France and England during the sixteenth century and other historians have drawn attention to the economic importance of Venice and the influence of this upon the neighbouring university of Padua. For a Marxist, indeed, this attractive analogy assumes the status of self-evident truth, according to which the mechanistic view of the universe reflects the mechanically dominated economy of the early modern period.

It is worth considering, however, that the machine was no new phenomenon in western Europe and the kind of machines with which the sixteenth century was familiar were not re-volutionary in design or concept. The most characteristic machines such as the windmill, the sailing ship and the wind pump used a source of power which had long been familiar to the West. The most radical machine was the cannon, a weapon of war, but hardly mechanical in its inspiration. In a sense the cross bow was more typically a machine, and this would bring us back to the thirteenth century. Nicolas Oresme (d. 1382) used scientific analogies based on the clock. In short, mechanical analogies were available to natural philosophers well before the end of the sixteenth century. What is needed is an explanation why at this time Galileo and his successors should seize upon a mechanical analogy as particularly appropriate.

The answer seems to lie in the revival of Archimedean science during the course of the sixteenth century. Archimedes (287-212 BC) was the greatest Greek mathematician. He was fascinated by mechanical analogies, as for example in his analysis of the lever, though the machines he conceived were designed for ornament and interest, not practical use. For

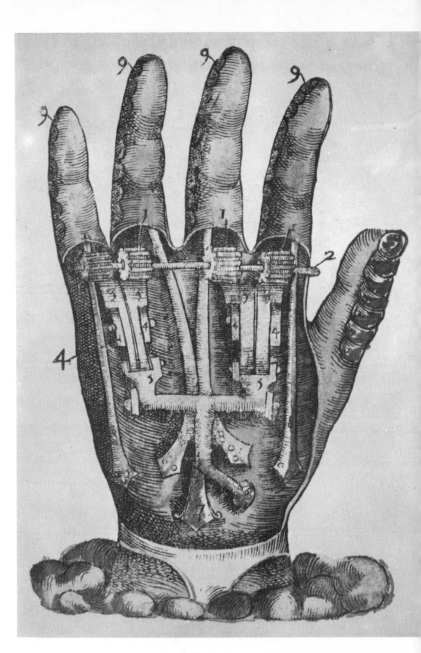

L'altro diſſe che molto piu tiraria a dui ponti piu baſſo di tal ſquare (laquale era diuiſa in 12.parte)come di ſotto appare in diſegno.

some sixteenth-century scientists, the works of Archimedes provided information about an aspect of Greek science which was neither Aristotelian nor Platonic.

There was an immense gap between the approach of the magical tradition and the detached intellectual curiosity of Archimedes. The Archimedean tradition was that of the mechanical engineer. It was not esoteric, it was not obsessed with the occult and it was not searching for mathematical

The sixteenth-century Italian engineer Niccolo Tartaglia (d. 1557) was one of the precursors of the mechanistic tradition in science. This drawing shows the application of geometric analysis to artillery projectiles.

harmonies of a religious significance. All this came as a novelty in the sixteenth century, though the works of Archimedes were known in manuscripts to a few medieval intellectuals in Latin translation. But the appearance of a printed edition in the mid sixteenth century was the real turning point.

The most important Archimedean of the early sixteenth century was Niccolo Tartaglia (1499-1557) who published the first Latin edition of Archimedes in 1543. This was followed by another version, Commandino's, published in 1575. Tartaglia himself was interested in problems with a practical bent, such as the trajectory of projectiles, a problem which had devastating implications for Aristotelian physics though Tartaglia himself did not raise them. Tartaglia was interested essentially in the principles of machines, an interest which was carried forward into Galileo's day by the *Liber Mechanicorum* (1577) of Guidobaldo, mathematician and patron of Galileo. The Archimedean revival, stretching from Tartaglia up to Galileo and beyond, laid the basis for a mathematical approach in which the world was open to measurement and analysis. In this tradition, numbers did not possess the mystical appeal that they did for the Platonists and neo-Platonists.

If mechanism was a reaction against magic, it was equally a reaction against the organic tradition. It was impossible to look upon the universe as a machine and to leave intact the existing Aristotelian assumptions about the nature of God, Christian revelation, miracles and the place of purpose in the world. The mechanist assumption was that the universe operated on the basis of mechanical forces, and, as Mersenne explicitly put it, God was the Great Engineer. Thus the task of the scientist was to explore the inter-relationship of the various parts of the universe, on the assumption that they would fit together like those of a machine.

These were then the three main intellectual traditions which are relevant to the study of Scientific Revolution, each with their own assumptions about God, Nature and scientific method. Inevitably the picture has been over-simplified. There were differences of emphasis within each tradition and there were variations in the course of time. Nevertheless, if we see the history of science in terms of three traditions, we will in some measure escape from the dangers of the Whig interpretation of the history of science. We are now more likely to think in terms of several scientific methods than of a single one. And we will certainly place less emphasis upon rational assumptions. Indeed, judged by rational criteria the magical tradition appears to be the least rational of the three; yet judged by its contribution to the Scientific Revolution, we may see it as the most important.

2 Scientific styles, languages and experiments

From another point of view, these traditions appear as sources of model-building or, what is much the same thing, as the grammars of specific languages. Within a particular tradition, the world of nature was interpreted according to characteristic models and concepts. There was no single 'scientific' paradigm to which the whole of western natural philosophy conformed. There were several paradigms, or traditions, which shaped the way in which men looked at and thought and talked about the world, and each of them was the possible source of insights. Whig interpreters of history have adopted the mechanistic tradition as the scientific tradition *tout court*, but to do this is to reduce a complex picture to a caricature. Modern science grew out of a conversation carried on in several competing languages. By this, I do not mean language in the narrow sense of a working vocabulary, but in the much wider sense of a complete mental approach. Language provided the 'eyes', through which men looked at nature and the source of whatever concepts and analogies were held to be appropriate in the exploration of nature.

Another feature of scientific style may be seen in the scientists' choice of certain experiments as especially instructive. It is generally held that one of the features of the Scientific Revolution was the rise of the experimental method. This is true but too simple. Experiment for its own sake is apt to be pointless; to make sense it must be made against the background of a general hypothesis. For this reason, it is illuminating to consider which experiments, or which types of experiment, were considered to be exciting or important within terms of a particular tradition.

Scientific languages

If we discuss first the language of Aristotelian science, it seems clear that this derived ultimately from Aristotelian metaphysical concepts about the structure of the universe. A metaphysical language, using such terms as 'substance' and 'accident', 'matter' and 'form', 'essence' and 'existence', was applied by Aristotelians to the world of nature. This laguage was taught within the seventeenth-century universities by using textbooks, of which Franz Burgerdyck's *Idea Philosophiae* may be taken as a typical example. Burgersdyck, who was professor at Leyden during the 1620s and 1630s, wrote a number of textbooks, all of them reprinted many times during the course of the century. In his discussion of physics he used the Aristotelian definition of movement (from potentiality to actuality) and referred his readers to Aristotle's *Physics* book III. In discussing the heavens, he raised such questions as whether they were corruptible or not and provided references to Aristotle's *Physics* and *Metaphysics* as well as to contemporary Jesuit textbooks. Perhaps most characteristic of all, Burgersdyck looked upon disciplined argument (the 'disputation') as the most acceptable way of studying physics. This was the language of the universities, Catholic and Protestant alike.

What was true of Protestant Leyden was equally true of Catholic Bologna. The young student at Bologna, even though he intended to take up medicine, began first with Aristotelian logic, physics and metaphysics and then by a natural progression took up Aristotle's scientific treatises, especially the *Meteorology*, *Generation and Corruption* and *Historia Animalium*, before going on to study Galen and Avicenna. A professor might alternate as a matter of course between lecturing on

This illustration is from Elias Ashmole *Theatrum Chemicum Britannicum*. The four figures are the great luminaries of the alchemical tradition with Hermes Trismegistus prominent among them. Ashmole himself (1617-92) – mathematician, alchemist, astrologer and editor of John Dee – was a prime figure in the magical tradition.

philosophy and lecturing on medicine. There was no sharp
division between medicine and metaphysics. In 1661 all can-
didates for a doctorate at Bologna had to swear allegiance to
Aristotle as well as to Galen and Hippocrates.

Even in the middle of the seventeenth century, Aristotelianism was still the dominant scientific language. Indeed, as we
have seen above, there had been a remarkable revival of
scholasticism, the significance of which is still underrated. It
was precisely the strength of Aristotelianism which accounts
for the missionary fervour of those who pressed the claims of
their own rival languages. There was a great deal of emotion
on all sides. It was not a scientific debate according to the
accepted stereotype of sober detached discussion but a war of
words, a battle of the books.

During the sixteenth century, the main alternative scientific
language to Aristotelianism was the language of magic and
alchemy – or rather the languages, since alchemy was a Tower
of Babel run by a medley of competing maguses. There was
no single theme in alchemy, but perhaps it is not too misleading
to regard the search for the philosopher's stone, which would
transmute base metals into gold, as the archetypal model of
most alchemical activity. Much of the language used sounds
fantastic to modern ears, but it was an attempt to introduce
some measure of order and control into the overwhelming
exuberance of nature.

Most alchemists dealt with what is in some sense the equivalent of modern chemistry. They prepared substances by various
methods, distillation for example, in the hope that they would
tap a source of miraculous power. But if a discovery was made,
the likelihood of it being passed on to a second generation
was diminished by the emphasis on secrecy.

The prime difficulty for alchemists was to decide upon the

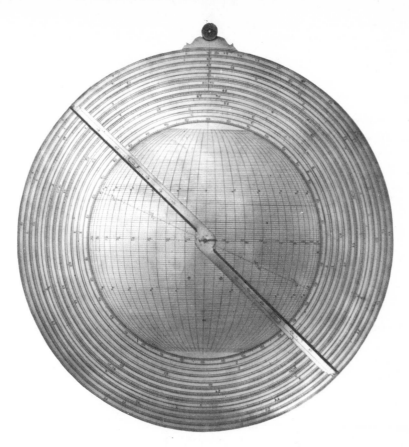

criteria which made one substance different from another. Almost inevitably they were led to depend upon qualitative differences. They regarded colour as particularly important in offering a clue to the fundamental properties of a substance. Black was associated with death, green with fruitfulness. Taste was also used to differentiate substances. Sugar of lead was the name given to what we call lead acetate. The mystic name 'vitriol' (Visita Interiora Terrae Rectificando Invenies Occultum Lapidem – Visit the Interior of the Earth and by Right Action You will Find the Occult Stone) was used as a description for substances shiny and crystalline. Other substances were given names which derived from their discoverers. For example Glauber's spirit of nitre took its name from Glauber the German chemist (1604-70).

Seven metals were associated with the sun, the moon and the five known planets. Lead for example was analogous to Saturn, the slowest moving and hence presumably the heaviest planet. These associations, in conjunction with the concepts of acid, alkali, salt and spirit provided a further source of nomenclature. Terms like 'spirit of Venus' or 'salt of Saturn' were still in current use in the eighteenth century. Other terms such as 'salammoniac' survive into the twentieth century.

These names, many of which seem merely picturesque to modern eyes, were intended to provide a clue to the chemical character of the substance in question. The difficulty was that there was no unifying element in the language used. But we should not identify disunity with chaos. Perhaps historical studies today with a mingling of formal and informal approaches may be used as a good modern analogy of the alchemical world of the sixteenth and seventeenth centuries.

There were a number of intellectuals in the seventeenth century who were conscious of the difficulty presented by the absence of appropriate scientific language. Prominent among them was Francis Bacon. In his *New Organon* Bacon distinguished four obstacles to the path to truth: the Idols of the Tribe, the Cave, the Market Place and the Theatre. Of these, the idols of the Market Place originated in language:

The idols imposed by words on the understanding are of two kinds. They are either names of things which do not exist . . . or they are names of things which exist, but yet confused and ill-defined, and hastily and irregularly derived from realities. (*New Organon*, part 2, Aphorism LX.)

Bacon attempted to create a new language of science, and hence a new system of classification. He described for example 'Prerogative Instances' (key observations) numbering twenty-seven in all. The best known of these is the 'Instance of the Finger Post' (originally 'instancia crucis' or Instance of the Cross) by which a single experiment could be set up to decide between contradictory hypotheses. Bacon also used the term 'form' by which he meant laws, e.g. laws of light or heat.

Bacon's attempt to introduce a new vocabulary was taken seriously by the Royal Society, but his specific suggestions were not in fact adopted. Robert Boyle in 1666 paid due deference to the Baconian divisions of Natural History but then went on to substitute his own:

I blame not the Verulamian division, and do much less pretend to propose a perfect one; yet I shall venture to substitute another as that which seems to me more suitable to the Immensity and variety of the Particulars that pertain to Natural History.[2]

Boyle's most famous book *The Sceptical Chemist* was an attack upon the language used by the Aristotelians and the alchemists, and he was particularly hard upon the latter for their 'obscure, ambiguous and almost enigmatical way of expressing what they have to teach'.

But the language which developed most during the seventeenth century was a language ignored by Bacon. It was mathematics. In the nature of things, its use was confined to a small number of people, though the number grew in the course of the seventeenth century. The technical development of mathematics, as it took place at the hands of the Frenchmen, Fermat, Pascal and Descartes, was analogous to the process by which a language acquires a more elaborate structure and range of expression and hence is able to deal

H/ COCK EXCVD / CVM PRIVILEGIO.

DEBENT   IGNARI · RES   FERRE ET POST OPERARI
IVS · LAPIDIS   CARI   VILIS SED DENIQ3 RARI
VNICA   RES · CERTA   VILIS SED VBIQ3 REPERTA

QVATVOR   INSERTA   NATVRIS   IN   .NVBE   REFERTA
NVLLA   MINERALIS   RES   EST   VBI   PRINCIPALIS
SED   TALIS   QVALIS   REPERITVR   VBIQ3   LOCALIS

with modifications of feeling and sensibility which were once beyond it.

This efflorescence of mathematics was comparable to the burst of activity in logic which took place during the twelfth-century Renaissance when, thanks to Peter Abelard and his successors, logic became the exciting new intellectual language, which could be applied to the whole range of experience. In the seventeenth century, a similar development took place in the field of mathematics, with Descartes playing the role of Abelard. Mathematics held an overwhelming appeal for almost all the original minds of the seventeenth century. It was a language which combined beauty and clarity to a unique degree and, what is more, it appeared to be providing the keys to the Universe. Those who learned the language, who expanded its potentialities and who pressed for its adoption were drawn largely from the ranks of the mechanists.

It is difficult to overemphasise the importance of mathematics within the mechanist tradition. Advances in mathematics culminating in the calculus, discovered by Newton and Leibniz, made possible an intellectual sophistication which outstripped all other forms of human reasoning. The new mathematics opened up new frontiers in mechanics, ballistics and astronomy and its very success in these fields implied that the world of nature was built on a mathematical model. For this reason as much as any other, mechanism was dominant at the end of the seventeenth century.

Style in experiment

In the organic tradition, Harvey's treatise on the circulation of the blood (1628) offers an appropriate starting point. Harvey described an experiment by Galen in which he tried to show that arteries are held to pulsate like bellows and hence provide a pumping action, though he rejected Galen's conclusions.

Harvey also referred to Galen's experiment upon the trachea of a living dog. On the basis of these and other experiments, he concluded that the cause of the 'perturbation' of the arteries was the heart. The arteries become distended, but they are filled like bladders, not because they expand and contract like a bellows. In order to prove this point, Harvey experimented on fishes, frogs, pigeons and snakes and claimed to have performed experiments on the heart which Galen and Vesalius had recommended but not themselves carried out. One experiment at least was carried out on the corpse of a man who had just been hanged.

Harvey also referred to Galen's experiment upon the ficant. He calculated, given the size of the heart, how much blood could be transferred into the arteries in a given time and tested his hypothesis by experiments, which included observation of the slower circulation in fishes and reptiles. Clearly, imaginative and precise experiment did not belong only to Bacon or the Royal Society.

Another form of research within the Aristotelian tradition was based upon the close and painstaking observation of nature. Marcello Malpighi (1628-94) professor at the university of Bologna for much of his life, may be taken as an appropriate instance here. Malpighi's classical work was in the field of embryology, where he examined the development of the embryo of the young chicken. The study of the embryo raised questions which could not be solved within a narrowly mechanistic framework and which were not raised by the alchemists, and hence this type of study arose more naturally within the Aristotelian tradition than in either of the two other scientific traditions.

The difference between the Aristotelian and mechanistic points of view may be illustrated from a simple contrast between Malpighi and the mechanist, Borelli (1608-79). Borelli was a product of the tradition of Galileo and Torricelli at Florence and published a study of anatomy in which mechanical analogies were used to explain the movement of the limbs. As a young man Malpighi had been influenced by Borelli but he eventually reacted against him and chose embryology as a field of research. In this field the Cartesian notion of the machine seemed to Malpighi to have little relevance. Aristotelian language seemed much more appropriate since the development of an embryo presupposed some kind of final cause.

Within the alchemical tradition, experimental method also had its own distinctive flavour. The alchemists' aim was to discover a fundamental base metal which by the application of the philosopher's stone could be transformed into gold. All this involved a great use of fire and furnaces as a tool of research.

The alchemical style of experiment emerges with particular force in the work of Van Helmont (1577-1644). Helmont never tired of urging his readers to test authorities by 'the touchstone of experiment'. But the kind of experiments which he had in mind rested upon metaphysical assumptions about the nature of the universe. In rejecting the Aristotelian belief in four basic elements, earth, air, fire and water, Helmont came to believe that water was the prime constituent of matter. He proved this by his famous tree experiment:

I took an earthen vessel, in which I put 200 pounds of earth that had been dried in a furnace, which I moystened with Rain-water, and I implanted therein the Trunk or Stem of a Willow Tree, weigh-

With the use of the magnifying glass, Marcello
Malpighi (1628-94) took Aristotelian studies of the
embryo of a chick a great step forward. These are
uncompromisingly 'scientific' drawings, compared
with those of Vesalius (see pages 82-3), which
combined a scientific and an aesthetic appearance.

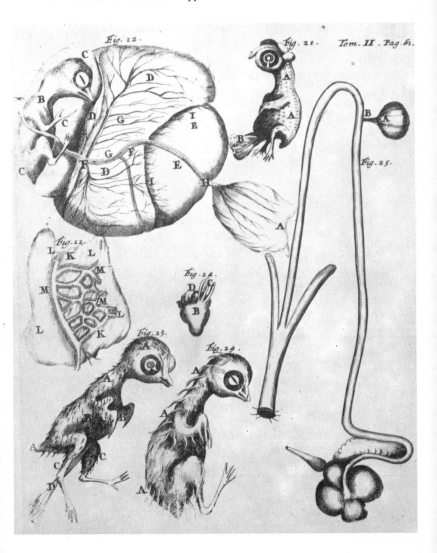

ing five pounds; and at length, five years being finished, the Tree sprung from thence did weigh 169 pounds and about three ounces. But I moystened the Earthen vessel with Rain-water, or distilled water (always when there was need), and it was large and implanted into the Earth, and lest the dust that flew about should be co-mingled with the Earth, I covered the lip and mouth of the Vessel . . . I computed not the weight of the leaves that fell off in the four Automnes. At length, I again dried the Earth of the Vessel, and there were found the same 200 pounds, wanting about two ounces. Therefore 164 pounds of Wood, Barks and Roots arose out of the water only.[3]

Helmont also believed that such natural processes as fermentation offered a better clue to the workings of nature than those provided by mechanical analogies. Hence he was led to develop the techniques of distillation to a point which they never reached in the two rival traditions. Helmont carried out this type of experiment meticulously, so much so that he received high praise from Robert Boyle.

Helmont was also led to experiment with gases, as a result of rejecting Aristotelian doctrine on the air. Hitherto, all forms of what we now regard as phenomena of gases had been explained in terms of one of the four elements – air. Helmont, by experiment, was able to show that there were various forms of air, which he labelled as 'gases'. This interest in often dangerous experimenting with gases rested upon Helmont's assumptions about the universe. 'Gas' for him took its name from the Greek word 'chaos'. It was not the neutral term which it is for us, but heavy with overtones about the structure of the universe. Helmont's experiments, admirable as they were, were part of his general style of thought and language.

Another experimenter in the alchemical tradition was Dr

Robert Plot (1640-97), professor of chemistry at Oxford and member of the Royal Society. In Plot's papers there is a copy of an agreement which he made to investigate the possibility of discovering a universal remedy. The document made explicit reference to the Hermetic philosophers including Basil Valentine and Paracelsus and it obviously envisaged experiments taking place in Plot's laboratory. Strict attention was to be paid to secrecy and the persons who signed the agreement were enjoined 'never to divulge either directly or indirectly anything of the matter or process to any person whatsoever, without the knowledge and consent of the author'. The aim of the experiments was set out as follows:

That which the Author proposeth to be performed on his part is as followeth:
First, He will undertake to shew and that not barely by notion, but by matter of fact, what the true matter of the Hermetick philosophers is . . . which is the very key which unlocks all their secrets, by the knowledge of which their choise Hieroglyphicks are unfolded, their emblems unmasked their dark riddles and Philosophical Parables laid open, yea their very hidden misteryes discovered . . .[4]

This document shows that the magical tradition was far from dead in the late seventeenth century. Its chief interest for us, however, is to demonstrate how the search for a magical elixir dictated a particular style of experiment.

Mechanistic style in experiment

The tone of experiment within the mechanistic tradition was set by Galileo by experiments which he described in his *Dialogues* and *Discourses*. Where an Aristotelian like Harvey dealt with the real behaviour of actual blood, Galileo moved

A German miner's compass and accessories of wood and ivory in its original wooden box, 157 × 136 mm. The compass box bears a pivotted hook at each end, so that it may be hung on a line. The compass is dated 1689, the curved scale, 1690. The printed sheet stuck on the underside of the lid of the box includes a perpetual almanac based on the Golden Number, and an astrological table.

as far as possible from day to day 'reality' by setting up experiments which provided a model for an abstract universe. Aristotelians regarded mathematics as an escape from the problems set by the real world of constant change; Galileo, on the other hand, invented a mathematical world in which speed, time and distance were the only considerations.

We may see this in an account which he gave of an experiment with an inclined plane:

SALV. The request which you, as a man of science, make, is a very reasonable one; for [such experiment] is the custom – and properly so – in those sciences which apply mathematical demonstrations to the study of natural phenomena, as is seen in the case of perspective, astronomy, mechanics, music, and others where the principles, once established by well-chosen experiments, become the foundations of the entire superstructure. I hope therefore it will not appear to be a waste of time if we discuss at considerable length this first and most fundamental question. I have attempted in the following manner to assure myself [about] the acceleration actually experienced by falling bodies.

A piece of wooden moudling or scantling, about 12 cubits long, half a cubit wide, and three finger-breadths thick, was taken; on its edge was cut a channel a little more than one finger in breadth; having made this groove very straight, smooth, and polished, and having lined it with parchment, also as smooth and polished as possible, we rolled along it a hard, smooth, and very round bronze ball. Having placed this board in a sloping position, by lifting one end some one or two cubits above the other, we rolled the ball, along the channel, noting, in a manner presently to be described, the time required to make the descent. We repeated this experiment more than once in order to measure the time with an accuracy such that the deviation between two observations never exceeded one-tenth of a pulse-beat. Having performed this operation and having assured ourselves of its reliability, we now rolled the ball only one-

quarter the length of the channel; and having measured the time of its descent, we found it precisely one-half of the former. Next we tried other distances, comparing the time for the whole length with that for the half, or with that for two-thirds, or three-fourths, or indeed for any fraction; in such experiments, repeated a full hundred times, we always found that the spaces traversed were to each other as the squares of the times, and this was true for all inclinations of the plane, i.e., of the channel, along which we rolled the ball. We also observed that the times of descent, for various inclinations of the plane, bore to one another precisely that ratio which, as we shall see later, the Author had predicted and demonstrated for them.

For the measurement of time, we employed a large vessel of water placed in an elevated position; to the bottom of this vessel was soldered a pipe of small diameter giving a thin jet of water, which we collected in a small glass during the time of each descent, whether for the whole length of the channel or for a part of its length; the water thus collected was weighed, after each descent, on a very accurate balance; the differences and ratios of these weights gave us the differences and ratios of the times, and this with such accuracy that although the operation was repeated many, many times, there was no appreciable discrepancy in the results.[5]

Several points are worth noticing about this experiment. In the first place, it was repeated a hundred times. Secondly, great pains were taken to achieve accuracy of measurement. Thirdly, conditions regarded as extraneous to the experiment such as friction were reduced to a minimum. In the conditions of his time, this was as near as Galileo could hope to reach towards creating an artificial environment in which the 'essential' factors were capable of being measured.

The object of this experiment was to elucidate the problem of acceleration raised by a falling body. In the same treatise Galileo also showed that 'if a body falls freely along smooth planes inclined at any angle whatsoever, but of the same height, the speeds with which it reaches the bottom are the same', and went on to prove that 'the spaces described by a body falling from rest with a uniformly accelerated motion, are to each other as the squares of the time intervals employed in traversing those distances'. This meant that the Aristotelian view that speed was proportional to weight was completely false. The key was the square of the time involved.

It was his analysis of the acceleration of a body, freely falling from rest, which gave Galileo the starting point from which to analyse the motion of a projectile. He was able to show that the speed on leaving the muzzle remained constant. Acceleration only occurred where gravity was involved.

Another experiment often performed by seventeenth-century mechanists was described by Galileo in the *Discourses*:

Aristotle says that 'an iron ball of one hundred pounds falling from a height of one hundred cubits reaches the ground before a one-pound ball has fallen a single cubit'. I [the Character Salviati] say that they arrive at the same time. You find on making the experiment, that the larger outstrips the smaller by two finger-breadths, that is

This diagram, from Galileo's *Dialogues concerning Two New Sciences* (1638), illustrates his mathematical approach to explaining the motion of falling bodies. Hitherto it had been assumed that the speed of a falling body was related either to its weight or the distance fallen. Galileo argued, first mathematically and then experimentally, that the speed of a falling body was related to the square of the time involved. Distance after accelerated motion is represented in this diagram by the area of the triangles ABC, AFI and APO where AC, AI and AO represent the times. The areas of the triangles are in ratios of 1, 4 and 9, i.e. the squares of the times 1, 2 and 3.

A Flemish planispheric astrolabe of gilt brass, diameter 287 mm., made in 1565 by Regnerus Arsenius, the nephew of Gemma Frisius (1508-55). A planispheric astrolabe is an instrument for solving astronomical problems on the positions of the sun and the stars (e.g. time-telling and times of rising and setting) and includes an alidade for measuring altitudes.

when the larger has reached the ground, the other is short of it by two finger breadths – Aristotle declares that bodies of different weights, in the same medium, travel . . . with speeds proportional to their weight . . . If you wish to maintain the general proposition you will have to show that the same ratio of speeds is preserved in the case of all heavy bodies and that a stone of twenty pounds moves ten times as rapidly as one of two; but I claim that this is false, and that if they fall from a height of fifty or a hundred cubits, they will reach the earth at the same moment.[6]

This experiment was not original. A thousand years before, the Byzantine scholar John Philoponus described a similar procedure in these words:

For if you let fall from the same height two weights of which one is many times as heavy as the other, you will see that the ratio of times required for the motion does not depend on the ratio of the weights but that the difference in time is a very small one. And so, if the difference in weights is not considerable, that is, if one is, let us say, double the other, there will be no difference, or else an imperceptible difference.[7]

Galileo's contemporary, the Dutch mathematician and engineer Simon Stevin (1548-1620), described the experiment in print in 1586:

The experience against Aristotle is the following. Let us take (as the very learned Mr Jan Cornets de Groot, most industrious investigator of the secrets of Nature and myself have done) two spheres of lead, the one ten times larger and heavier than the other, and drop them together from a height of 30 feet onto a board or something on which they will give a perceptible sound. Then it will be found that the lighter will not be ten times longer on its way than the heavier but that they will fall together onto the board so simultaneously that their two sounds seem to be one and the same rap.[8]

Thus this experiment of Galileo was not new. What was new was that they became crucial issues in a battle between two paradigms, the organic and the mechanistic. Hence they were performed over and over again.

It would be misleading however to emphasise the role of experiment in Galileo's methods at the expense of mathematics. The impression given by the *Discourses* is of an overwhelmingly mathematical treatise, based largely upon Euclid, with experiment playing a minor part. Galileo was willing to devise an experiment from which he excluded all considerations but space, time and distance. But such an experiment was as much illustration as observation. The real significance of his approach was the assumption that the universe operated according to the laws of mechanics.

This brings us to a further point about Galileo and the mechanist tradition. The accounts of experiments published by Galileo tended to be polemical in character, for his aim was to destroy the Aristotelian tradition in physics and astronomy. This outlook greatly influenced the mechanist tradition during the course of the century. The mechanists concentrated upon experiments which exposed the weaknesses of the organic tradition and the zeal in overthrowing orthodoxy gave the writings a certain missionary fervour. It is only fair to say that by the nineteenth century, mechanism had itself acquired some of the intolerant characteristics of an orthodoxy.

The need to destroy the basis of Aristotelian physics also explains the mechanists' interest amounting almost to obsession with experiments designed to demonstrate the possibility of a vacuum. The vacuum was as much of a battleground between the organic and mechanist traditions as the behaviour of projectiles. For Aristotelians, belief in a vacuum was the criterion of a philosophy which derived final causes in the

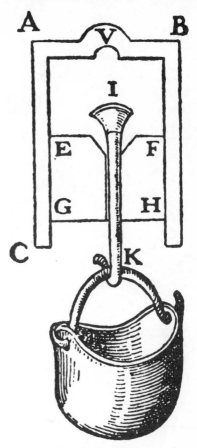

Galileo's experiment to prove the existence of a vacuum. The vacuum provided a test case for deciding against Aristotelianism. It was the seventeenth-century equivalent of the 'missing link' as a debating point.

universe and was bound up with the atomism of Democritus and Epicurus. To admit the possibility of a vacuum meant admitting that atomism was a valid hypothesis. Thus the Aristotelians would go to any lengths to deny or disprove the point. (In this they were at one with the Galenists, since Galen in his writings made a particular point of attacking the Epicureans for their belief that the human body, and its organs, was the product of chance not design.)

A brief statement of the Aristotelian doctrine on the vacuum is to be found in Galileo's *Discourses* when Simplicio, the spokesman for the Aristotelian point of view, stated:

Left An *equatorium* of brass, diameter 193 mm., for calculating planetary longitudes. This example is probably French. *Equatoria* in metal are very rare, although more common in manuscripts and printed books. *Right* Night dials or nocturnals were made from about the fifteenth century for converting the time indicated by the position of a star into solar time. The most easily recognisable stars were chosen for this purpose: for instance, the two bright stars of Ursa major, easily discerned by anyone, provide a very convenient pointer on the celestial clock. *Bottom left* Graphometre of gilt brass (1597). The graphometre remained the basic surveying instrument until the close of the eighteenth century.

Bottom right A German astronomical *compendium* of gilt brass, diameter 72 mm., made in Augsburg in 1588. The top and bottom lids are hinged and open to reveal various instruments. Starting from the upper side of the top lid, the instruments included in this *compendium* are: an astrolabe-quadrant, a wind-vane, a map of part of Europe with a pivotted index and a cursor, a horizontal dial adjustable for use in several latitudes, and a compass, a table of latitudes of 48 towns, another type of sundial, a lunar volvelle and *aspectarium*, and scales showing the length of day and night and times of sunrise and sunset.

Second air pump of Robert Boyle
(1627-91). This is an example of a machine
constructed to create a vacuum as a basis
for scientific experiment. Boyle's scientific
curiosity was an expensive hobby.

Aristotle inveighs against the ancient view that a vacuum is a necessary prerequisite for motion and that the latter could not occur without the former. In opposition to this view, Aristotle shows that it is precisely the phenomenon of motion which renders untenable the idea of a vacuum.[9]

The emphasis which the Aristotelians placed upon the impossibility of a vacuum, provided a challenge for mechanists to set up an experiment which showed a vacuum was possible. Torricelli claimed to have proved the existence of a vacuum by a simple experiment in which he inverted a tube full of mercury in a bowl of mercury. He claimed that the space left at the top of the tube was a vacuum. This phenomenon was known to sixteenth-century scientists using tubes of water but this was the first time that such a conclusion had been drawn from it (i.e. it had now been proved in a mechanistic context). Torricelli's experiment, like Galileo's inclined plane experiment, acquired much of its significance from its value as a weapon against Aristotelianism and for much of the seventeenth century it retained a symbolic value.

Inventions

The role of the invention in the Scientific Revolution may be discussed against a background of scientific styles and languages. Inventions were developed not in the general atmosphere of 'inventiveness' but within a particular tradition. It was no accident for example that mechanists like Galileo, Hooke and Huygens should have devoted considerable attention to the clock, since within the mechanist tradition both the precise measurement of time and the significance of the clock counted for more than in the rival traditions. A similar judgment may

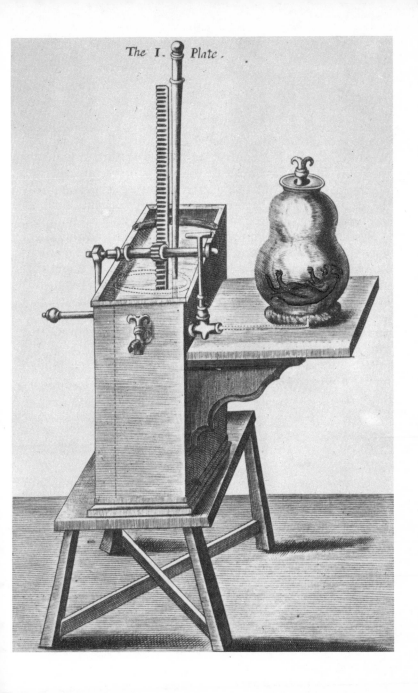

The I. Plate.

be made of Boyle's development of the air-pump, which was essential for his vacuum experiments. Within a different tradition Tycho developed instruments for ascertaining the precise position of the planets for his own purposes, namely the drawing up of horoscopes. Glauber's invention of an improved furnace falls into place within the magical tradition.

By placing inventions within a particular tradition, light may also be thrown upon the vexed question of the relationship between science and technology. There is no doubt that inventions, such as the telescope, were made by craftsmen outside the world of learning. Gilbert, for example, learnt a great deal from the observations of 'mechanics'. But the crucial point, to my mind, is that such discoveries were incorporated into a particular scientific tradition. Thus Galileo saw the possibilities of the telescope for his own purposes, even though he did not invent it.

3 The world as organism

The contribution of Padua

Within the organic tradition, the main contribution to the Scientific Revolution was made by professors at the University of Padua, and by its most celebrated English alumnus William Harvey. At Padua, an influential tradition of 'scientific' scholasticism had survived throughout the fourteenth and fifteenth centuries. This tradition had originated in the universities of Oxford and Paris during the fourteenth century and then spread to Italy. At Oxford, the great names were those of the Merton College group, Thomas Bradwardine (1290-1349), and his pupils Richard Swineshead and William Heytesbury (1380). At Paris the leading figures were Jean Buridan (1300-60) and Nicole Oresme (+ 1382). At Padua the great physician Jacopo da Forli (+ 1461) wrote commentaries on the works of Nicole Oresme.

The intellectual father of much of this approach was the English philosopher, William of Ockham (+ 1349), whom we may describe not unfairly as a fourteenth-century Bertrand Russell. His successors at Oxford and Paris (often called Terminists because they used a logical method technically known as *a terminis*) followed Ockham in concentrating upon particular problems and disregarding the construction of great systems. According to Ockham's Razor, entities were not to be invented unnecessarily.

The most celebrated achievement of the school was to offer an alternative hypothesis to the accepted Aristotelian explanation of motion. Orthodox Aristotelians, such as Aquinas, attributed the continuing motion of a projectile to the effect of the medium through which it was passing, in this case the air. This explanation did not satisfy the Terminists and they developed a theory in which the local motion of a projectile

The anatomy theatre at Padua, capital of the organic
tradition in science. In February 1646 the diarist John Evelyn
attended 'the famous Anatomic lecture, which' he wrote
'is here celebrated with extraordinary apparatus,
and lasting almost the whole month, in which I saw
three, a woman, a child and a man dissected'.

or falling body was attributed to a new quality ('impetus')
acquired by the body. As a hypothesis this had the advantage
of leading the Terminists on to a more fruitful range of problems and historians of science are agreed in looking upon their
work as an imaginative step forward within the Aristotelian
tradition. It has been argued recently, for example, that
Bradwardine, working on 'impetus' premises, anticipated
Galileo in proving that all bodies fell at an equal speed within
a void.

Interest in this range of problems survived at Padua in the
sixteenth century, after it had died out at Oxford and Paris.
The Paduan counterpart of the Oxford and Paris Terminists
was Giambattista Benedetti (1530-90) who used the impetus
theory to explain the acceleration of falling bodies. Benedetti
also criticised the orthodox Aristotelian view that the velocity
of a falling body was proportional to its weight. The last
important representative of this school was Cesare Cremonini
(1550-1631).

This mathematical approach to nature developed within the
philosophical faculty at Padua, but the reputation of the
university rested upon its medical faculty, the most famous in
Europe. It was the medical school of Padua which concentrated
upon the empirical side of Aristotelianism, most clearly revealed in the *Historia Animalium*. At Padua Giacomo Zabarella
(1532-89) was the best-known exponent of Aristotelian
empiricism.

If Aristotle the empiricist was one major influence upon
Padua, Galen was the other. The rise of the school of medicine
at Padua to European eminence was bound up with the revival of Galen in the fifteenth century. Jacopo da Forli,
professor first of medicine, then of natural philosophy,
founded a Galenist school which continued to flourish in

the sixteenth and early seventeenth centuries, thanks to the discovery of new treatises of Galen. The most famous professor in the sixteenth century was Andreas Vesalius (1514-64), 'father of modern anatomy', who was professor of surgery and anatomy at Padua for seven years (1537-44). Vesalius' famous anatomical treatise *De Fabrica* was published in 1543 while he was at Padua and the title page showed him carrying out a dissection there. His pupil Gabriele Fallopio (1523-63), discoverer of the Fallopian tubes, became professor in 1551. The tradition was carried on by Fabricius of Acquapendente (1537-1619), who was professor for fifty years. The anatomy theatre was built during this period (1595). There was thus an unbroken Galen tradition at Padua from the fifteenth century and it was at the feet of Fabricius that William Harvey (1578-1657) sat during his years as a student at Padua.

Harvey and the circulation of the blood

Of scientists working within the Galenic-Aristotelian tradition, William Harvey made the most decisive contribution to the Scientific Revolution by his discovery of the circulation of the blood. After his years at Padua, Harvey returned to England, where in 1607 he was elected fellow of the Royal College of Physicians. He was appointed lecturer in anatomy to that conservative body in 1615 and his lectures followed duly acceptable lines. Harvey seems to have kept in the main to the Paduan tradition, though he also used an Aristotelian textbook originating in Basel.

When he came to lecture on the heart, he was faced with a choice between the teaching of Galen or Aristotle. According to Galen, the heart was merely one of three 'principles' of the body, the other two being the liver and the brain. Aristotle, on the other hand, regarded the heart as the main source of blood in the body. Both authors saw the heart as analogous to a spring, from which blood was distributed throughout the body (though in Galen's case, it was one of three springs). Harvey came to the conclusion that blood circulated through the heart some time between 1616, when he first gave the lectures, and 1628, when he published his book *De Motu Cordis*.

Harvey seems to have reached this conclusion in two stages. First, he moved away from Galen, not towards a 'modern', as we might surmise, but to Aristotle, whom he followed in the view that the heart was the source of blood in the body (not the only instance in which he chose Aristotle as his guide). From this Aristotelian starting point, Harvey, in a second stage of reasoning and experiment, concluded that blood did not originate in the heart but passed through it. Another, and

Overleaf The illustrations for Vesalius' great work *De Fabrica* (1543) were drawn by a pupil of Titian. They show how, in the sixteenth century, there was no sharp division between arts and science. (Contrast the drawings for Descartes' *Tractatus de Homine* pages 161-2). The skeleton (*right*) is more than a medical illustration. It evokes in its stance and in the position of the skull a sixteenth-century moral attitude, possibly of defiance towards death.

equally important point is that Harvey did not draw the conclusion that the heart was a pump. He regarded it as a means by which the venous, impoverished blood was restored to its full warmth and nutritive content. Harvey did not see the importance of the lesser circulation through the lungs. He held to the assumption, inherent in Aristotle, that the heart was an organ of immense significance, not merely a pump.

Though Harvey followed Aristotle in this instance his general debt to Galen was considerable. Historians have tended to regard Galen as an 'authority' who blocked Harvey's path to further advance but it has been convincingly argued that this is a prejudice based upon a misunderstanding about the nature of Galen's teaching on the heart. It overlooks, for example, Galen's accurate description of the four sets of cardiac valves. Galen may have failed to make sense of the minor circulation of the blood through the lungs (the pulmonary circulation) but his actual anatomical description did not act as a positive barrier in the way of Harvey's imaginative leap.

In other words, Harvey belonged to a medical tradition which provided him with a foundation for further advance. The common interpretation of this episode in the history of science is to trace a line of anti-Aristotelian and anti-Galenist revolts from Vesalius through Servetus to Fabricius of Aquapendente, a rebellius lineage which enabled Harvey to dethrone Galen. But modern historians have shown conclusively that this interpretation is mistaken and rests upon fundamental misconceptions about Galen's teaching. Harvey referred to Galen as 'that great prince of physicians' and he relied upon Galen's accurate description of the cardiac valves. Harvey himself implied that Galen provided the evidence but did not draw the conclusions.

PRIMA
MVSCVLO-
RVM TA-
BVLA.

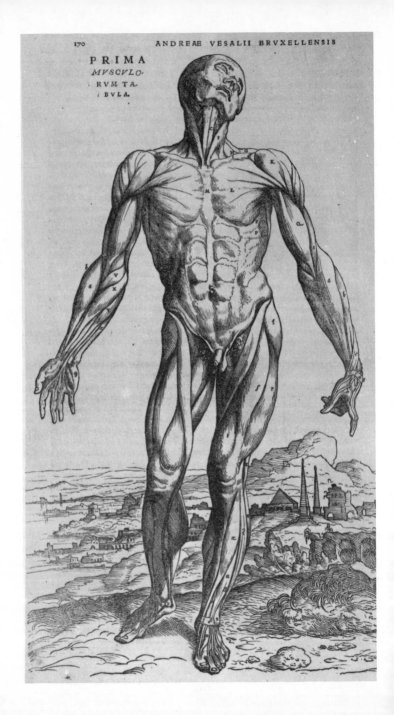

HVMANI COR-
SIMVL COMPACTO-
EX FACIE EXPRES-

PORIS OSSIVM
RVM ANTERIORI
SIO.

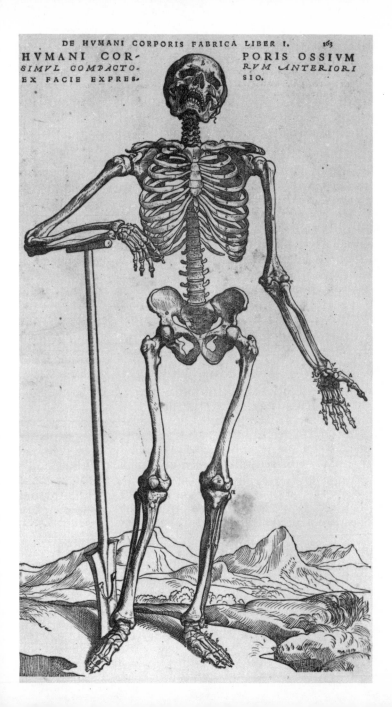

An illustration from William Harvey's *De Motu Cordis*, in which he proved the circulation of the blood.

Harvey's own account of his discovery of the circulation of the blood is very revealing:

I began to think whether there might not be a motion as it were in a circle. Now this I afterwards found to be true ... which motion we may be allowed to call circular, in the same way as Aristotle says that the air and the rain emulate the circular motion of the superior bodies; for the moist earth, warmed by the sun, evaporates; the vapours drawn upwards are condensed, and descending in the form of rain moisten the earth again; and by this arrangement are generations of living things produced; and in like manner too are tempests and meteors engendered by the circular motion, and by the approach and recession of the sun. And so in all likelihood does it come to pass in the body through the motion of the blood; the various parts are nourished, cherished, quickened by the warmer more perfect vaporous spiritous and as I may say alimentive blood; which, on the contrary, in contact with these parts, becomes cooled, coagulated and so to speak effete; whence it returns to its sovereign, the heart, as if to its source, or to the inmost home of the body, there to recover its state of excellence or perfection. Here it resumes its due fluidity and receives an infusion of natural heat – powerful, fervid, a kind of treasury of life, and is impregnated with spirits, and it might be said, with the balsam, and thence it is again dispersed; and all this depends on the motion and action of the heart. The heart, consequently, is the beginning of life; the sun of the microcosm, even as the sun in its turn might well be described as the heart of the world; for it is the heart . . . which . . . .is indeed the foundation of life, the source of all action.[10]

This passage throws a great deal of light upon Harvey's manner of reasoning. It is clear, first of all, that his imagination was not struck by a mechanical analogy. The heart is not a pump, it is the sovereign of the blood and its inmost home, where its excellence is restored. The analogies here are political (the heart as the monarch) and domestic (the heart as the

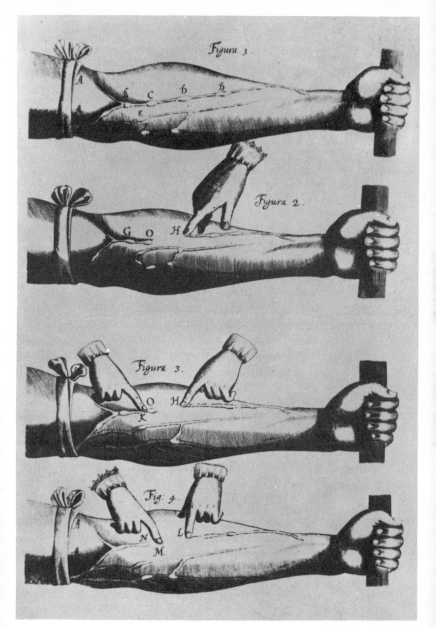

restoring fireside). The heart is also the treasury, providing the blood with a refurbished currency. Harvey also draws an analogy from the weather and the way in which there is a recurring cycle of vapour and moisture. But this was not a mechanical cycle. Harvey saw it, through Aristotle's eyes, as 'the arrangement by which generations of living things are produced'. It is, in short, part of the Aristotelian world picture, in which final causes and organic growth and decay were regarded as the key to understanding nature.

Harvey's use of circular motion as an appropriate analogy was derived from his Aristotelian background and education in the university of Padua. Aristotle saw circular motion as the only perfect form of movement and hence appropriate to the planets. In transferring this to the microcosm, Harvey was using an Aristotelian analogy, although Aristotle would have regarded it as inappropriate within the sublunar setting. Here again his mode of thought is thoroughly Aristotelian. It seems fair to say that Harvey developed his ideas on the circulation of the blood within the general framework of the 'old philosophy'. Later Descartes transformed the notion of circulation into a mechanical one and used it as a cornerstone for his mechanical philosophy.

Elsewhere Harvey stressed the place of final cause in the world and attacked those who followed Epicurus and Lucretius in attributing natural effects to chance:

They who philosophise in this way, assign a material cause (for generation) and deduce the causes of natural things either from the elements concurring spontaneously or accidentally, or from atoms variously arranged; they do not attain to that which is first in the operations of nature and in the generation and nutrition of animals; viz. they do not recognise that efficient cause and divinity of nature which works at all times with consummate art, and providence and wisdom, and ever for a certain purpose, and to some good end; they derogate from the honour of the Divine Architect, who has not contrived the shell for the defence of the egg with less of skill and foresight than he has composed all the other parts . . .[11]

This passage brings out well the appeal which Aristotelianism made to the defenders of orthodoxy, during a period when scepticism appeared to be on the increase. It shows us Harvey attacking the atomists precisely because atomism was associated with the Democritan world governed by chance. It also explains why Harvey should see human blood as carrying something over and above material nourishment since it was also the vehicle of a vital principle, that is, the Aristotelian soul. On all points we see Harvey making a religious stand as much as a scientific discovery and drawing his inspiration from the organic tradition.

It would be misleading to conclude this discussion of the organic tradition without stressing the point that 'scientific scholasticism' was an exceptional phenomenon in the late sixteenth and early seventeenth centuries. The tides of religious orthodoxy were more favourable to Aquinas than to Ockham, to the system builders of the thirteenth century than to their fourteenth-century critics. Scientific scholasticism survived at Padua only because the university was situated within the Republic of Venice, a tolerant community by the standards of Loyola and Calvin. Elsewhere a type of scholasticism triumphed in which theological considerations were paramount.

What this scholastic revival meant in practice may be seen in the career of Robert Bellarmine (1542-1621), who presided in 1616 over the Papal Commission which condemned the proposition that the sun is the centre of the world. In his early years, Bellarmine was sent from Italy to inaugurate the theological course at the Jesuit house of Louvain in 1570, where he

The title page of Francis Bacon's *Instauratio Magna* (1620). Lord Chancellor Bacon (1561-1626) saw himself as sailing to an intellectual new world through the Pillars of Hercules. Harvey thought he wrote science like a Lord Chancellor.

displayed an overwhelming enthusiasm for Aquinas. Indeed he appeared to prefer Aquinas before any other guide including the scriptures themselves. In his lectures at this time Bellarmine wrote:

He [Thomas] expounds everything in so beautiful an order in a manner so easy and concise, that if one studies with care these few questions of St Thomas I dare to state categorically that he will find nothing difficult in all that touches the Trinity in the Scriptures, the Councils and the Fathers, and in persevering with the study of the holy doctor he will make more progress in two months than if he devotes a great number to a direct and personal study of the Scriptures and the Fathers.[12]

The Scientific Revolution did not emerge from this type of scholastic thinking but from a sceptical tradition which went back to Ockham and ultimately to Aristotle, the empirical observer.

Francis Bacon – a neo-Aristotelian?

The name of Francis Bacon (1561-1626) can scarcely be omitted from any account of the Scientific Revolution. Bacon, an astonishing mixture of lawyer, philosopher, scientist, politician and moralist, influenced seventeenth-century thought in many ways, not least in the inspiration which his writings provided for the Royal Society. If we look ahead to the generation which followed him, it is clear that his name was adopted by the mechanists as a support for their cause. Robert Hooke, the leading exponent of mechanism in the Royal Society (see page 178) envisaged a scientific method based on mechanical assumptions about nature and he attributed the origins of his 'Philosophical Algebra', as he said, to

FRANCISCI

DE VERULAMIO,

Summi Angliæ

CANCELARIJ,

Instauratio

magna.

Multi pertransibunt & augebitur scientia.

Sim: Pass: sculp:

Anno

LONDINI
Apud Joannem Billium
Typographum
Regium.

1620.

'the incomparable Verulam'. But if we turn to Bacon himself, the mechanistic content of his thought is minimal.

Bacon certainly attacked Aristotelianism and saw himself as correcting its fault of excessive rationalism. 'Research into final causes', he wrote, 'like a virgin dedicated to God is barren and produces nothing'. He also criticised the unguided empirical enthusiasm of Paracelsus and alchemists generally. From this, it is possible to conclude that his outlook was mechanistic, in embryo at least, and, as support, to point to his belief that heat was 'true mechanical motion'.

But, over the whole range of his writings, Bacon adopts a point of view which links him more with the organic tradition than any other. If this seems paradoxical, the explanation lies in the fact that Bacon created a misleading model of Aristotelianism, which implied that Aristotelians did not experiment. As we have seen, there was an Aristotelian tradition of experimenting and Bacon's ideas about the role of experiment derive from this rather than anywhere else.

Bacon called upon men

to sell their books and to build furnaces; quitting and forsaking Minerva and the Muses as barren virgins and relying upon Vulcan. (*The Advancement of Learning*, Book I)

He considered it certain

that unto the deep, fruitful and operative study of many sciences, specially natural philosophy and physics, books be not only the instrumentals ... In general, there will be hardly any main proficience in the disclosing of nature, except for expenses about experiments. (*Ibid.*)

In his general approach to science Bacon criticised the narrow empiricist and the dogmatic theorist, looking ahead to a combination of experiment and theory.

The men of experiment are like the ant; they only collect and use; the reasoners resemble spiders, who make cobwebs out of their own substance. But the bee takes a middle course; it gathers its material from the flowers of the garden and of the field, but transforms and digests it by a power of its own. (*New Organon*, Part II, XCV)

He provided a concrete instance of what he had in mind in a utopian treatise, *The New Atlantis*, where research was to be conducted by thirty-six scientists within an establishment called 'Solomon's House'. There were large caves to be used for research into refrigeration, metals and the curing of diseases, high towers for observing meteors, and weather, and great lakes, fountains, walls, orchards, and parks each with their own research interests. Bacon also described brew-houses and kitchens, dispensaries, furnaces, perspective houses and sound houses (for problems of light and sound) and engine houses for imitating various kinds of motion. He refers to 'a mathematical-house where are represented all instruments, as well of geometry as astronomy, exquisitely made'.

All this was very much in the spirit of encyclopedic accumulation of knowledge and dependent upon Aristotelian concepts such as 'humours', 'natural' and 'violent' motion. Where the mechanists applied a single mechanistic method to natural phenomena, Bacon advocated observations over a wide range of phenomena, as in the following list of topics which he drew up for investigation. In this approach he recalls Aristotle far more than any scientist in the mechanist tradition:

History of Species
28 History of Fossils; as Vitriol, Sulphur, etc.
29 History of Gems; as the Diamond, the Ruby, etc.
30 History of Stones; as Marble, Touchstone, Flint, etc.

31 History of the Magnet.

32 History of Miscellaneous Bodies, which are neither entirely Fossil nor Vegetable; as Salts, Amber, Ambergris, etc.

33 Chemical History of Metals and Minerals.

Next come histories of man

45 History of Humours in Man; Blood, Bile, Seed, etc.

46 History of Excrements; Spittle, Urine, Sweats, Stools, Hair of the Head, Hairs of the Body, Whitlows, Nails and the like.

47 History of Faculties; Attraction, Digestion, Retention, Expulsion, Sanguification, Assimilation of Aliment into the members, conversion of Blood and Flower of Blood into Spirit, etc.

48 History of Natural and Involuntary Motions; as Motion of the Heart, the Pulses, Sneezing, Lungs, Erection, etc.

49 History of Motions partly Natural and partly Violent; as of Respiration, Cough, Urine, Stool, etc.

50 History of Voluntary Motions; as of the Instruments of Articulation of Words; Motions of the Eyes, Tongue, Jaws, Hands, Fingers; of Swallowing, etc.

51 History of Sleep and Dreams.

52 History of different habits of Body – Fat, Lean; of the Complexions (as they call them), etc.

71 History of Smell and Smells.

72 History of Taste and Tastes.

73 History of Touch, and the objects of Touch.

74 History of Venus, as a species of Touch.

75 History of Bodily Pains, as species of Touch.

76 History of Pleasure and Pain in general.

77 History of the Affections; as Anger, Love, Shame, etc.

78 History of the Intellectual Faculties; Reflexion, Imagination, Discourse, Memory, etc.

79 History of Natural Divinations.

80 History of Diagnostics, or Secret Natural Judgments.

81 History of Cookery, and the arts thereto belonging, as of the Butcher, Poulterer, etc.

100 History of working in Iron.
101 History of Stone-cutting.
102 History of the making of Bricks and Tiles.
103 History of Pottery.
104 History of Cements, etc.
105 History of working in Wood.
106 History of working in Lead.
107 History of Glass and all vitreous substances, and of Glass-making.
108 History of Architecture generally.
109 History of Waggons, Chariots, Litters, etc.
110 History of Printings, of Books, of Writing, of Sealing; of Ink, Pen, Paper, Parchments, etc.

Bacon's organic approach is also shown in his account of the following experiment (in his *New Organon*):

Of all substances that we are acquainted with, the one which most readily receives and loses heat is air; as is best seen in calendar glasses [air thermoscopes], which are made thus. Take a glass with a hollow belly, a thin and oblong neck; turn it upside down and lower it, with the mouth downwards and the belly upwards, into another glass vessel containing water; and let the mouth of the inserted vessel touch the bottom of the receiving vessel, and its neck lean slightly against the mouth of the other, so that it can stand. And that this may be done more conveniently, apply a little wax to the mouth of the receiving glass, but not so as to seal its mouth quite up; in order that the motion, of which we are going to speak, and which is very facile and delicate, may not be impeded by want of a supply of air.

The lowered glass, before being inserted into the other, must be heated before a fire in its upper part, that is its belly. Now when it is placed in the position I have described, the air which was dilated by the heat will, after a lapse of time sufficient to allow for the extinction of that adventitious heat, withdraw and contract itself to the same extension or dimension as that of the surrounding air at the time of the immersion of the glass; and will draw the water upwards to a corresponding height. To the side of the glass there should be affixed a strip of paper, narrow and oblong, and marked with as many degrees as you choose. You will then see, according as the day is warm or cold, that the air contracts under the action of cold, and expands under the action of heat; as will be seen by the water rising when the air contracts, and sinking when it dilates. But the air's sense of heat and cold is so subtle and exquisite as far to exceed the perception of the human touch, insomuch that a ray of sunshine, or the heat of the breath, much more the heat of one's hand placed on the top of the glass, will cause the water immediately to sink in a perceptible degree. And yet I think that animal spirits have a sense of heat and cold more exquisite still, were it not that it is impeded and deadened by the grossness of the body. (*New Organon*, Second Book of Aphorisms, XIII, 38)

The interest of this experiment is twofold. First, though Bacon does not mention the fact, it had been performed or seen by others. Secondly, Bacon refers to air as having 'a sense of heat and cold so subtle and exquisite as far to exceed the perception of the human touch'. In other words, he uses a human analogy, not a mechanical one, to describe a natural phenomenon, and this brings his assumptions within the organic tradition.

The difficulty of assessing Bacon's contribution to the Scientific Revolution is bedevilled in part by the fact that he exaggerated the extent of his own originality by concealing his sources. Another cause of confusion lies in the use which was made of his name by the mechanists of the next generation and even by millenarian thinkers like Comenius. At the present moment there remains a considerable doubt about his own contribution, and it is not clear whether he stands there in his own right or merely because others thought him important in the seventeenth century. Some historians emphasise the magical overtones of Baconian thought. The debate continues, with more sympathy for his achievements in England than elsewhere. We may perhaps conclude that his real contribution lay in popularising the notion of experiment, even though his own ideas on experimental technique were vague in the extreme.

4 The mysterious universe

Copernicus

During the sixteenth century, the tradition of magic and art introduced a distinctive dimension to science which may be seen in the growing importance attached to mathematics, astrology, astronomy and chemical analysis. It was from within this tradition that the first radical criticisms were made of the geocentric theory of the universe, which centuries of Aristotelian dominance had turned into orthodoxy. During the fifteenth century, the voice of Nicholas of Cusa (1400-64), cardinal and theologian, had been raised in the Platonist cause, but the first decisive steps were taken by the astronomer Nicholas Copernicus (1473-1543).

The history of modern science is normally taken to begin with Copernicus and the grounds for the judgment are substantial. Copernicus rationalised the Ptolemaic cosmology by placing the sun at the centre of the universe, with the earth moving round it in an annual revolution. In addition to this, Copernicus argued that the earth rotated on its axis every twenty-four hours. Thus the difference between day and night was no longer explained on the basis of a cosmology, according to which the sun and the planets revolved around the earth each day.

This was certainly a revolutionary step, but Copernicus did more than assume it as a general proposition. He carried its implications into a re-assessment of the mass of astronomical observations which were available in Ptolemy's *Almagest*. His achievement was an extraordinarily imaginative detailed working out of original assumptions, performed by a mathematician of very high competence. In short it was not simply a poetic vision, though it was that; it was also a piece of technical mathematics.

In many respects, Copernicus seems a most unlikely person to have taken such a radical step. He was born in the frontier regions between Germany and Poland at Torun (in German, Thorn), where urbanisation had made little progress and where, we might assume, the pursuit of scholarship was unlikely to flourish. His nationality was Polish, or German, or both, a problem which gave rise to difficulty during the Second World War when the fourth centenary of his book was celebrated (1943). Copernicus himself was the nephew of a bishop and, though he himself never became a priest, he enjoyed ecclesiastical preferment which enabled him to follow a life of leisure. He was, in fact, a gentleman, albeit one of recent origin.

In astronomy, Copernicus found the leisured occupation par excellence. He himself made few astronomical observations. He was content to deal with mathematical calculations on the basis of observations made by others, and in this we may see him as thoroughly within the Greek intellectual tradition, uninterested in the practical world. From 1512 until his death in 1543 Copernicus was canon of the cathedral at Frauenburg. He had no ecclesiastical duties and he was able to follow a life of secluded study, in contrast to the semi-military activities of his fellow canons. Copernicus was a classic case of the isolated intellectual, cut off from society, enclosed within the walls of the cathedral grounds, remote from the pursuits of his fellows and part of a social unit which was in its turn isolated from the society around it.

Copernicus's devotion to astronomy appears against this background as a form of retreat from the demands of the 'real' world. He was not a professor and he never became one. Astronomy of this kind was a form of scientific asceticism, a sixteenth-century equivalent of the life of the hermit. (Indeed

Pythagoreans like the Florentine Toscanelli were literally ascetics.)

But his isolation in Poland is not in itself sufficient to account for his revolutionary outlook. The explanation for this must be sought in the ten years which he spent in Renaissance Italy. After four years' student life at the university of Cracow, Copernicus left for further study at the universities of Bologna and Padua. It was in Italy that he came under the influence of neo-Platonism, an influence revealed in a letter attributed to Pythagoras which he translated and which stated: 'it is not permitted to let ordinary men into the sacred mysteries of the Elysian goddesses'.[13]

This and other pieces of evidence show that the Renaissance by which Copernicus was influenced was the Renaissance of Ficino and the Florentine academy. Perhaps we should see his introduction to neo-Platonism as the equivalent of a religious conversion.

The link with neo-Platonism was provided by Copernicus's associate and teacher at Bologna, Domenico Maria de Novara who knew the Florentine neo-Platonists and who translated Proclus and Hermes Trismegistus. Proclus (AD 412-85) attributed a mystical value to mathematics:

The soul [of the world] therefore, is by no means to be compared to a smooth tablet, void of all reasons, but she is an ever-written tablet, herself inscribing the characters in herself, of which she derives an eternal plenitude from intellect . . . All mathematical species, therefore, have a primary subsistence in the soul, so that before sensible members there are to be found in her inmost recesses, self moving members; vital figures, prior to the apparent; ideal proportions of harmony . . . here we must follow the doctrine of Timaeus, who derives the origin and consummates the fabric of the soul, from mathematical forms . . .[14]

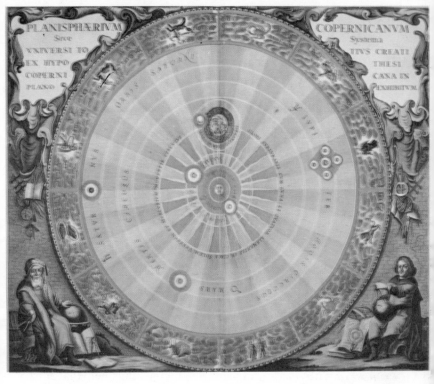

Neo-Platonic emphasis upon the sun may also be seen in a quotation from Marsilio Ficino:

Nothing more reveals the nature of the Good (which is God) more fully than the light (of the sun). First light is the most brilliant and clearest of sensible objects. Second, there is nothing which spreads out so easily, broadly, or rapidly as light . . .
Just look at the skies, I pray you, citizens of the heavenly father-land . . . The sun can signify God himself to you, and who shall dare to say that the sun is false.[15]

Finally, a quotation from Copernicus which explicitly mentions Trismegistus:

In the middle of all sits the Sun enthroned. In this most beautiful temple could we place this luminary in any better position from which he can illuminate the whole at once? He is rightly called the Lamp, the Mind, the Ruler of the Universe; Hermes Trismegistus names him the visible God, Sophocles Electra calls him the All-seeing. So the sun sits as upon a royal throne ruling his children the planets which circle around him.[16]

The link between modern science and the Renaissance has often been discerned. Butterfield, among others, stressed the importance of the close natural observation which Leonardo undertook. Others have drawn attention to the mathematical expertise which was involved in an artistic use of perspective. But the kind of Renaissance influence implied in Proclus, Trismegistus and Ficino is of a very different order. Neo-Platonism flourished during the later years of Cosimo de Medici and Lorenzo de Medici. It involved a turning away from involvement in the 'real world' either of politics or of art. The kind of masterpiece which it fostered was Botticelli's *Primavera* with its aura of symbolic magic. This neo-Platonist tradition led towards a mystical reverence for numbers, not a wholesome respect for practical mathematical techniques. It encouraged secrecy and an interest in the occult for its own sake by which a work of art was seen as a magical emblem or a coded message for the initiate. This Renaissance was a world away from the rationalism of Machiavelli, and its full secrets, thanks to the work of Wind and others, are only now being revealed.[17]

This was the background of Copernicus. Indeed Rheticus thought that Copernicus delayed publication of his work to preserve its secret for the favoured few, so that: 'the Pythagorean principle would be observed, that philosophy must be pursued in such a way that its inner secrets are reserved for learned men trained in mathematics'.[18]

The neo-Platonic background of Copernicus also explains why his theories were almost universally rejected through the sixteenth century. Only the neo-Platonists accepted Copernicus without reserve. Edward Rosen has listed those of the religious establishment who spoke out against the apparently absurd notion that the earth went round the sun. The reaction to Copernicus from the world of orthodoxy, both Catholic and Protestant, was hostile. Even before Copernicus's treatise had been published, Luther reacted violently to rumours about it. Luther, who was nothing if not biblically minded, said:

That is how things go nowadays. Anyone who wants to be clever must not let himself like what others do. He must produce his own product as this man does, who wishes to turn the whole of astronomy upside down. But I believe in the Holy Scripture, since Joshua ordered the sun, not the earth, to stand still.[19]

Melanchthon, who was much more of an Aristotelian than Luther, made similar criticisms ten years later, in 1549:

Out of love for novelty or in order to make a show of cleverness, some people have argued that the earth moves. They maintain that neither the eighth sphere nor the sun moves. Whereas they attribute motion to the other celestial spheres, and also place the sun among the heavenly bodies. Nor were these jokes invented recently. There is still extant Archimedes' book on *The Sandreckoner* in which he reports that Aristarchus of Samos propounded the paradox that the sun stands still and the earth revolves around the sun.

Even though subtle experts institute many investigations for the sake of exercising their ingenuity, nevertheless public proclamation of absurd opinions in indecent and sets a harmful example.[20]

Professor Rosen lists several other examples in the late sixteenth century, including Robert Recorde, author of *The Castle of Knowledge*, the standard textbook on astronomy in

'The most famous and most learned doctor, Nicolaus Copernicus, incomparable astronomer'.

CLARISSIMUS · ET · DOCTISSIMUS · DOC
TOR · NICOLAUS · COPERNICUS · TORU
NENSIS · CANONICUS · WARMIENSIS
ASTRONOMUS · INCOMPARABILIS 1875

R. H. Tawney described Luther's reactions to the economic changes of the early sixteenth century as those of a savage examining a watch which he could not comprehend. Luther's attitude to the theories of Copernicus was much the same.

England. Scaliger and Buchanan both held anti-Copernican views and Jean Bodin, so often cited in textbooks of political thought as a modern, argued that it was impossible for a simple body like the earth to move in the three different ways which Copernicus assigned to it. The same point was made by the most eminent astronomer of the late sixteenth century, Tycho Brahe (1546-1601), when he wrote:

What need is there without any justification to imagine the earth, a dark, dense and inert mass, to be a heavenly body undergoing even more numerous revolutions than the others, that is to say, subject to a triple motion, in negation not only of all physical truth, but also of the authority of the Holy Scripture which ought to be paramount.[21]

The truth is that within the general range of religious opinion, Catholic, Lutheran and Calvinist alike, the Copernican view was dismissed as an absurdity. All the accepted authorities were against it. The Bible contradicted it expressly. The weight of common sense acted as an additional obstacle. Equally powerful was the fact that there seemed to be no way of proving it. During the sixteenth century the heliocentric view was accepted only within the Pythagorean-Hermetic tradition. On Hermetic assumptions the central place of the sun in the universe seemed axiomatic because it was 'fitting'. On Aristotelian assumptions, the earth was the central point of the universe for exactly the same reasons. And since Aristotelianism was so strongly entrenched in the universities, it was inevitable that Copernicus's views should be rejected in the academic textbooks. Not until Galileo published his *Sidereus Nuntius* (1609), followed up by the *Letter to the Archduchess Christina* (1612) and the *Dialogues* (1632), did Copernicus obtain a champion outside the Hermetic tradition.

Bruno

The intellectual tradition which we have associated with Copernicus survived among a coterie, in the margins of the established educational and religious institutions of the age. It lost ground before the revival of Aristotelianism which followed the Council of Trent. It was attacked on the Protestant side by Thomas Erastus (1524-83). It suffered from the policies of intolerance in the late sixteenth century. In the world of art, the neo-Platonic tradition died the death. There was no Botticelli in the Italy dominated by Philip II. The Florentine Academy ceased to exist. Nevertheless, though it was under pressure, neo-Platonism survived.

The three men who were associated with Italian neo-Platonism during the period 1550-1600 were Francesco Patrizzi (1529-97), Giordano Bruno (1548-1600), and Tomaso Campanella (1568-1639). In 1591 Patrizzi published a large collection of the Hermetic writings with a dedication to Gregory XIV in which he appealed to the Pope to encourage the teaching of Plato and of Platonists like Plotinus, Proclus and the early church Fathers. He himself had lectured on neo-Platonism at the University of Ferrara, but Ferrara was a centre of minor academic importance and Patrizzi's appeal met with only a temporary and limited response and his book was condemned as heretical. He died in his bed in 1597, but the fate of his book was symptomatic of the general intellectual climate of Italy. The tide was running hard in favour of Aristotelianism and a philosopher who pressed the claims of heliocentricism was bound to run into trouble.

The outstanding Italian exponent of the magical tradition was Giordano Bruno, born at Nola, near Naples in 1548, who became a Dominican in 1563. His intellectual precocity led to charges of heresy and he went into exile. He adopted the life of a wandering scholar, and his life recalls that of Paracelsus whom indeed he admired (see page 114). The difference between the two men may be explained by the publication of Copernicus's treatise *De Revolutionibus* in 1543, the year before Paracelsus died. Paracelsus was unaffected by Copernicanism, Bruno was dominated by it.

Bruno was the most enthusiastic exponent of the heliocentric doctrine in the second half of the century. He lectured throughout Europe on it and in his hands Copernicanism became part of the Hermetic tradition. Copernicus had hinted at this, but Bruno drew out the implications, and more, of Copernicus's reference to Trismegistus. Sixteenth-century reaction to the heliocentric doctrine cannot be fully understood without realising that heliocentrism remained within the Hermetic tradition until Galileo. Bruno transformed a mathematical synthesis into a religious doctrine.

Bruno saw the universe in the same terms as Lull, Ficino, and Pico had done, as a magical universe in which the earth and the stars were alive. Above all, the sun was alive, providing its light as a source of life at the centre of the universe. The task of the philosopher was to make use of the invisible forces which pervaded the universe, a task in which Trismegistus provided the essential key.

Inevitably these views brought him into conflict with the orthodox academics. The most famous of these controversies was his visit to Oxford in 1583 when he lectured to the dons upon the theory of Copernicus and 'many other matters'. Bruno has had historians of science on his side for the most part, for in spite of his eccentricities he seemed to represent 'rationalism'. But a recent discovery places beyond all doubt the fact that for him Copernicanism was part of a new intellectual system which derived in large measure from Trismegistus and the neo-Platonists. George Abbot in a work published in 1604 described how Bruno

... The Italian Didapper ... got up into the highest place of our best and most renowned schools, stripping up his sleeves like some jugler, and telling us much of theritrum and chirculus and circumferenthia (after the pronunciation of his Country language) he undertooke among very many other matters to set on foote the opinion of Copernicus that the earth did goe round, and the heavens did stand still; whereas in truth it was his own head which rather did run round and his brains did not stand still. When he had read his first lecture, a grave man, and both then and nowe of good place in that University, seemed to himself, somewhere to have read those things which the Doctor propounded: but silencing his conceit till he heard him the second time, remembered himself then, and repairing to his study, found both the former and later lecture taken almost verbatim out of the workes of Marsilius Ficinus.[22]

This passage tells us as much about Oxford as it does about Bruno. There is more than a hint of irrationalism about the description of the Italian Didapper and his odd, non-Oxonian pronunciation of Latin. The passage also reveals why Bruno's stand on Copernicanism should carry such little weight. It was based upon the pre-Copernican judgments of the Italian humanist Ficino, not upon any new arguments or observations. Not least interesting is the fact that Bruno's critic George Abbot had Puritan sympathies, which makes any close correlation of Puritanism and Science seem too simple. Perhaps Aristotelian logic stood here for rational thinking as against mystical enthusiasm.

Gilbert and magnetism

Bruno's precise contribution to science remains a matter for conjecture, but with his contemporary, the physician William Gilbert (1540-1603), there is no doubt. Gilbert's *De Magnete*, published in 1600, but written *c.* 1580, is the first substantial scientific treatise in English history, as well as being a land-mark in the Scientific Revolution. Gilbert's originality lay in undertaking a study of magnetism, a phenomenon which had been known at least since the age of the Greeks, and producing a new theory about its nature, based upon a series of precise and carefully recorded experiments.

Gilbert conducted about fifty experiments to illustrate the nature of magnetism. Much of what he had to say was well known to seamen from common observation of the magnetic compass, for example the fact that the compass needle varied considerably in its declination from the north. Gilbert sought an explanation for this phenomenon after first ridiculing the ideas of recent commentators. Gilbert's main point was that the erratic changes in the behaviour of the needle indicated that local variations in the earth's crust were responsible and not a constant cause such as the stars suggested by Ficino and others. He proved his point by describing an experiment with a spheri-cal lodestone (magnetic iron oxide) which was 'crumbled away at a part of its surface and so had a depression comparable to the Atlantic sea or great ocean'. Several needles were then laid on the lodestone and it was observed that variation occurred at the borderline between the sound areas of the stone and the decayed parts. Gilbert saw in this an appropriate analogy to that which happened on earth.

Gilbert's originality in devising experiments should not be exaggerated. For example, he took over from his contemporary

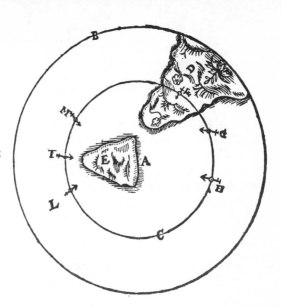

Diagram by William Gilbert to illustrate the behaviour of a magnet at different positions around the north pole of the earth, from *De Magnete* (1600).

Robert Norman an experiment in which a needle was pushed through a cork and made to float just below the surface in a glass of water. The object of this experiment was to disprove the theory that the basis of magnetism was attraction, since in fact the needle remained stationary in the water. Moreover, the sixteenth-century writers whom Gilbert attacked most, the Italians Porta and Cardano, believed in experiments as much as he did.

The interest of Gilbert lies as much in the formulation of his general theory of magnetism as in his experimental technique, though perhaps the distinction is an artificial one in view of the fact that the experiments were devised to lead to a general conclusion. Gilbert's theory of magnetism rested upon his judgment that the earth itself was a gigantic lodestone, and that 'every separate fragment of the earth exhibits in indubitable experiments the whole impetus of magnetic matter'. He held that the five main magnetic phenomena: attraction (which he called 'coition'), direction towards the earth's pole, variation, declination or dip and finally circular movement, could only be explained in terms of the earth's own magnetism.

The fifth aspect of magnetism, circular movement, provides a clue to Gilbert's views on the motion of the earth. He was not an avowed Copernican but he accepted the diurnal motion of the earth and proved it to his own satisfaction from the behaviour of lodestones. He thought that the lodestone, of all familiar objects, resembled the essential qualities of the earth in its capacity for 'whirling' motion.

Gilbert's attack on authority, his belief in experiment, his critical approach to evidence in constructing a general theory and his acceptance of the diurnal motion of the earth, all point to his modernity, and in a Whig interpretation of the history of science this is the conclusion which would be reached. A closer look at the *De Magnete*, however, will lead to second thoughts. For all his criticism of his predecessors, Gilbert found it impossible to avoid using much of their terminology and despite his condemnation of 'occult causes', his own explanation of magnetism remains fundamentally mysterious.

In fact, Gilbert's views fall into place as part of the 'magical tradition'. His experimental methods seem modern to us but his world outlook and his scientific assumptions are remote from the mechanism of the modern scientist. A quotation from *De Magnete* in which Gilbert specifically mentions Hermes Trismegistus serves to illustrate this point:

Aristotle's world would seem to be a monstrous creation, in which all things are perfect, vigorous and animate, while the earth alone, luckless small fraction, is imperfect, dead, inanimate and subject to decay. On the other hand, Hermes, Zoroaster, Orpheus recognise a universal soul. As for us, we deem the whole world animate, and all globes, all stars and this glorious earth too, we hold to be from the beginning by their own destinate souls governed and from them to have the impulse of self-preservation …

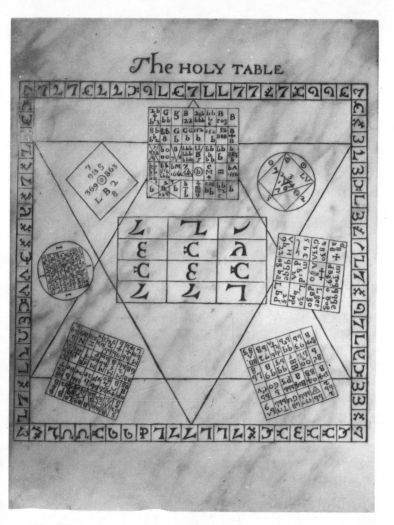

The HOLY TABLE

... Pitiable is the state of the stars, abject the lot of earth, if this high dignity of soul is denied to them, while it is granted to the worm, the ant, the roach, to plants ... [23]

He goes on to say that there are in the stars

reason, knowledge, science, judgment, whence proceed acts positive and definite from the very foundations and beginnings of the world.[24]

And again:

Wherefore not without reason Thales, as Aristotle reports in his book *De Anima*, declares the lodestone to be animate, a part of the animate mother earth and her beloved offspring.[25]

This Hermetic background led him to appreciate the virtues of Copernicus whom he described as 'the restorer of astronomy'. Gilbert referred to the 'primum mobile' of Aristotelian astronomy as 'inadmissible', 'this fiction', 'this product of imagination and mathematical hypothesis'.

This puts in brief many of the assumptions which we have ascribed to the magical tradition, notably that the world was animate, a belief which separated Gilbert from the Aristotelians, as well as from the mechanists. But it did not make him less of a scientist. Indeed we may argue that it was precisely his belief in the earth soul which led him to seek an explanation for magnetic phenomena which within the Aristotelian paradigm remained merely a curiosity.

From this point of view Gilbert was very much the contemporary of that fascinating and mysterious figure John Dee (1527-1608) who spanned the world of 'magic' and of 'science'. It is in fact impossible to separate Dee's magical interests from what we would now regard as 'legitimate' scientific ones. Like Gilbert he turned to Trismegistus and wrote in his diary of taking

Ghostly Council of Doctor Hannibal the great divine that had now set out some of his commentaries upon Pymander Hermetis Trismegisti.[26]

A similar kind of interest was to be seen in the scientific activity of the so-called 'Wizard Earl', Henry Percy, ninth Earl of Northumberland, who was imprisoned in the Tower of London from 1605 to 1621. Percy's place in the magical tradition is referred to in a poem by George Peele in which he was described as:

> Leaving our Schoolmen's vulgar trodden paths
> And following the ancient revered steps
> Of Trismegistry and Pythagoras
> Through uncouth ways and unaccessible
> Dost pass into the spacious pleasant fields
> Of divine science and philosophy.[27]

It is also illustrated by the fact that his library included Bruno's treatise on Lull (1588) and several treatises on alchemy, as well as Baptista Porta's *Natural Magic* (1585). Percy was not an isolated figure but the patron of gifted mathematicians such as Thomas Hariot (1560-1621).

Another member of the same group was Sir Walter Raleigh (1552-1618), who, like the Wizard Earl, was imprisoned in the Tower for many years. Raleigh's interest in chemical experiments has long been known. More recently, attention has been drawn to the emphasis he gave to alchemy and natural magic in his *History of the World* (1614). Raleigh praised four kinds of magic: divine magic, prophecy, astrology and chemical magic. Of this last, he wrote:

The third kind of Magick containeth the whole philosophy of Nature; not the babblings of the Aristotelians but that which bringeth to light the inmost virtues, and draweth them out of Nature's hidden bosome to humane use.[28]

Among the names he mentioned were Hermes, Ramon Lull, and Francesco Patrizzi, the Italian Platonist. Raleigh and Bacon wrote at the same time but they represent two distinct traditions in science, the one neo-Platonist and magical, the other neo-Aristotelian and anti-magical.

The title page of Raleigh's *History of the World* (1614) in which he accepted the teaching of Hermes Trismegistus as genuine. The title page has a suitably moral tone, though some historians have tried to read 'experientia' as 'experiment'.

Paracelsus

The figures of Copernicus and Paracelsus are not normally linked and superficially have little in common, but if we do bring them together we are led to appreciate the strength of the 'magical tradition' in the sixteenth century. Indeed it is only by bringing Copernicus and Paracelsus together that we may see the real shape of science during the period. If we do not do so, we will inevitably exaggerate the rationalism and the unity of the Scientific Revolution.

Paracelsus was a German doctor, born in Switzerland (1493-1547). Like Copernicus, he had some connection with the land-owning class, but his life took its shape from his sympathy with the lower social groups of Germany during this period, with the miners and the peasantry particularly. We may see him as a man with Anabaptist sympathies, taking his stand with the oppressed. This inevitably brought him into conflict with the social conservatism of Lutheran and Catholic orthodoxy and of the German universities. The failure of Paracelsus to make a substantial impact on the academic medicine of the day was because he was felt to be subversive. He was anti-intellectual in the sense of being anti-academic.

Paracelsus took as his main target the medical teaching of the universities, which drew its inspiration from Hippocrates and Galen and which in addition catered largely for the medical needs of the upper social groups. Physicians who qualified in the universities took the landowners and the bourgeoisie for their clientèle. The peasantry and the labourers sought help from the self-taught apothecaries who relied upon traditional rule of thumb and their own intuitions. It is difficult to decide who was worse off.

Academic physic rested upon the assumptions derived from

Raymond Lull (d. 1315), 'the enlightened doctor', developed 'the Lullian method' which enjoyed a revival in the sixteenth century. This diagram of the 'apostolic tree' formed part of a unified system of knowledge. His ideas form part of a tradition which, in its belief in mechanical aids to the acquisition of knowledge, stretched as far as Bacon and Comenius, and beyond.

two thousand years of Greek tradition that disease in the body derived from disorder in the balance of the four humours – phlegm, choler, melancholy and blood. This imbalance was known as 'distemper', hence all diseases were in a sense 'distemper'. The imbalance of the humours was thought to affect the body as a whole, not a particular part of it. Hence treatment was carried out with reference to the whole body, by attempting to redress the balance of the humours by means of bloodletting or induced vomiting or sweating. This was Greek medicine, still dominant in sixteenth-century Europe among those who were able to study it and to afford its benefits.

(This picture is no doubt over-simplified. Perhaps a distinction should be made between the influence of Hippocrates and that of Galen, with the Hippocratic tradition stressing the need for natural rest in achieving a cure, but it was Galen's influence which increased in the Renaissance period.)

Paracelsus attacked Galenic orthodoxy because of his social antagonism to a privileged elite. But there was also an intellectual foundation behind his radicalism. He drew inspiration from most of the anti-Aristotelian movements of the day, and not least from the neo-Platonism of Ficino. Paracelsus took the neo-Platonic doctrine of macrocosm and microcosm and applied it to the world of medicine. He saw the organs of the human body as the equivalent of the stars. The human body itself was a microcosm of all that existed in the world – animal, vegetable and mineral and spirit. The task of the doctor was to bring remedies from the macrocosm to cure the diseases of the microcosm.

In a sense, this looked forward to modern developments much more than did Galenic doctrine. But to make this judgment is to lose sight of the intellectual background in which Paracelsus had his origins. The world of Paracelsus

was the world of the Spanish eccentric Ramon Lull (1232-1315), Nicholas of Cusa, Pico della Mirandola (1463-94) and Ficino. It was a harking back to the pantheistic universe of Plotinus in which natural substances contained 'virtues' which were eternal in character and part of the divine nature. The universe was a magical universe in which God was the Magus. It was a world full of hidden secrets (the occult) which it was the task of the physician to overhear or to 'tune into'. It was a world dominated by spirit, not by matter, and hence completely different from the mechanistic world of Descartes and Hobbes. But it was not so very far from the worlds of Copernicus, and Kepler or of Bruno and Fludd. The place of Paracelsus in this tradition is quite clear.

All this sounds as strange to modern ears as the Galenic doctrine of the humours. But it did lead Paracelsus to concentrate his attention upon local areas of the body – the liver for example – rather than the body as a whole. He saw disease as coming from outside the body, not the result of humoral upsets, and he looked for specific remedies for specific diseases rather than resorting to generalised treatment like bloodletting. The Paracelsan cure for dropsy was mercury:

For example, Mercury is the specific remedy in dropsy. This is due to a morbid extraction of salt from the flesh, a chemical process of solution and coagulation. As such this process does not depend at all upon quality and complexion, but is a 'celestial virtue' endowed with its own 'monarchy' to which quality and complexion are subservient. Mercury will drive out the dissolved salt, which has a harmful corrosive action on the organs, and preserve the solid – coagulated – state of the salt in the flesh, where it is needed to prevent putrefaction. Mercury will effect the cure specifically in everybody, although it causes vomiting in one and sweating in another. Neither vomiting nor sweating – the universal cures of the ancients – are therefore the curative factors. Hence he errs who says the patient must be cured with sweating or vomiting, for he fails to consider the manifold variety of man and that any effect of such remedies is merely the expression of the different reaction of individuals to the same remedy, not the cure itself.[29]

Above all perhaps, Paracelsus gave a fillip to the study of chemistry within a disciplined framework. He went beyond the four traditional elements of Greek science (earth, air, fire and water) to three principles: sulphur, salt and mercury. These were not substances in the modern sense, but principles of activity, by which Paracelsus provided a new paradigm for the study of medicine and of 'chemistry'. He pointed the way for a search for new chemical remedies and a new emphasis upon experiment. The science of iatro-chemistry, in which chemistry was studied for use in medicine, was born.

Where he differed from Copernicus and Kepler was in his attitude to mathematics. He did not regard the universe as being written in mathematical characters. But in a sense this is a marginal difference. All these neo-Platonists believed in the same kind of universe and were all searching for a code which would reveal its secrets. Paracelsus differed from Copernicus in turning to the laboratories of the earth, the mines particularly, rather than the celestial laboratory.

Thus Paracelsus provides a link with the magical and mystical side of the Renaissance, which was once, and perhaps still is, grossly underrated. He looks back to the medieval world of Ramon Lull and forward to the seventeenth-century chemist, John Baptist van Helmont (1577-1644). If his achievements have not been appreciated it is because historians of science have always stressed the rationality of their pursuit of knowledge. But if Paracelsus was a lunatic so also were Nicholas of Cusa, Copernicus, Kepler and even Newton.

We have stressed the importance of Paracelsus's magical, religious and social assumptions in providing him with a basis from which he could attack Galenic orthodoxy. But we must not conclude too rashly that his emphasis upon localised centres of disease or the use of chemical medicines was an

The hazards of physic and surgery in the seventeenth century. From the title page of *De efficaci medicina* (1646).

Aduersa
fortiter

Chirurgia

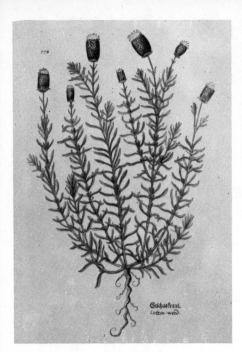

Left Cottonwell from *De historia stirpium commentarii* (1542) by Leonard Fuchs (1501–66). In this, as in other sixteenth-century engravings, art and science are inextricably intertwined.

Right 'The grete herball which geveth parfyt knowledge and un[der]standing of all maner of herbes and there gracyous vertues ... London 1526.' Herbals were descriptions of plants thought useful for medical purposes. The best known herbal was that of Dioscorides (first century AD) which appeared in print in 1478.

unmixed blessing. Paracelsus's emphasis in local causes may have blinded his followers to the essential unity of the human organism which modern medicine is at pains to emphasise and his use of chemical medicaments may have set back the clock as much as it moved it forward. In short, there were no ancients and no moderns in our sense of the word.

Paracelsus derived many of his ideas from another source impregnated with neo-Platonic ideas – the Cabbala. The Jewish Cabbala (literally 'tradition') was a body of teaching, dating from the later Middle Ages, which interpreted the Old Testament by esoteric methods, including cyphers. The revival of Hebrew among Christian scholars like Reuchlin (1455-1522) made the Cabbala accessible to those who sought wisdom behind the literal text of the Scriptures. The Cabbala and the Hermetic writings appealed to the same cast of mind. The Cambridge neo-Platonists for example studied the Cabbala.

Dē grotē herbari met al sijn figueren

der Cruyden / om die crachten der cruy
den te onderkennen. Met een tafele vanden namen der cruyden in latijn
en duytsche ❧ Een Regilter om lichtelic te vindē die curacien tegen alderhāde crāchede.
❧ Een tractaet om alle Brijnen te iudicerene. ❧ Wanden cantelen der Prüinen pā Meester
Arnoldus de noua villa / op dat die Meester niet bedroghen en worde. ❧ Om die operacien
vā allē drogerien eñ medecijnē te kennē. Metter Anothonije der menschelijcker ghebeeyten.
❧ Een expert Tractaet voor personē / die op dorpen / eñ kalteelen woonen / verre vā die
Meelters / Om te makene Wondtrāchē Saluē eñ oken / daer hem elck
met ghenesen mach. Welck tractaet in die ander Herbarius niet en is.
❧ Eñ noch veel ander nootlakelijcke leeringhen / Weder verduct int Iaer
M.CCCC. ende XXXII.

❧ Gheprint Tantwerpen / Bi mi Simon Cock.

The atmosphere of alchemy is successfully
conveyed here. Note the reference to Vitriol,
the cabalistic signs and the apparatus
to be found in an alchemist's laboratory.

The Paracelsan universe, in so far as it was neo-Platonic, was
alive. The three elements of Paracelsus were live sources of
spiritual energy analogous to the Trinity, which indeed they
'reflected'. All things had a spiritual directing force (an
Archeus) which guided their development.

What would be body without spirit? Absolutely nothing. The spirit
then and not the body contains concealed in itself the virtue and
power . . . [it] conserves the living body and when this perishes the
spirit escapes and leaves the dead body and returns to the place
whence it came, into the chaos, the air below and above the firma-
ment . . . know that spirit is the true life and balsam of all corporal
things.

Behind his insistence upon chemical remedies lay the assumption that he was living in a magical universe in which
material substances rightly chosen could achieve remarkable,
indeed miraculous, effects. Thus it was a magical not a 'rational'
assumption which led Paracelsus to bring about an alliance
between chemistry and medicine.

Before Paracelsus the world of alchemy was largely bound
up with the search for the philosopher's stone, which would
turn base metals into gold. The chances of success in this
venture were remote, to say the least, but the degree of skill
required in the search was considerable. The alchemist employed many of the techniques which were used in practical
trades, such as distilling and glass blowing. (The analogy here
is with Galileo, who took over a practical instrument, the
telescope, and used it to explore the heavens.)

There seems little doubt that alchemy revived during the
Renaissance. Classical fables were re-interpreted as sources of
alchemical knowledge. The Tale of the Golden Fleece, to
take one example, provided obvious inspiration for an alchemist seeking the philosopher's stone. But the nature of

alchemy tended to preclude any large-scale development. The alchemist concealed the technical secrets of his success and wrapped up his ideas in language of intentional obscurity in order to put his competitors off the scent.

Against this background, the achievement of Paracelsus seems to lie in bringing alchemy into the clearer world of iatrochemistry and of medicine. Relatively speaking, Paracelsus stood for an open, less secretive approach in alchemy. He wished to spread the gospel of the three elements, not to hide his light under a bushel. His language was obscure but it was not deliberately so; it lay in the nature of his ideas. But the tradition of secrecy in the alchemical tradition was a long time a-dying. Rudolph Glauber, the seventeenth-century chemist, described in a deliberately perverse way this method of producing *sal mirabile* (sodium sulphate), though his motives in doing so, the preservation of a patent and hence of profit, were 'rational' by modern standards.

Van Helmont

The influence of Paracelsus certainly increased during the first hundred years after his death, particularly among the apothecaries, who were concerned with the preparation of remedies. Since apothecaries were regarded as socially inferior to physicians, the advocacy of Paracelsan iatrochemistry possessed overtones of social radicalism.

In these circumstances it is odd to find that the most significant Paracelsan of the seventeenth century was a man of noble birth, Jean Baptista van Helmont (1577-1644). Helmont lived in the Spanish Netherlands, where aristocratic values were dominant but his views were not characteristic of his social environment, which he reacted against at an early age. For much of his life he was closely watched by the ecclesiastical authorities. His Paracelsan beliefs, it seems, were regarded as socially subversive as well as politically dangerous.

As a student, Helmont turned away from the scholasticism of his teachers to study the Cabbala, mysticism and magic. He took the degree of MD at Louvain at the age of thirty in 1609 after a period of travel which led at one stage to an offer of employment from Kepler's imperial patron, Rudolph II. He was never free from the attentions of the Inquisition from 1621 onwards. Helmont claimed to be orthodox and willing to submit to the Church over such matters as Copernicanism, but there seems little doubt that his Paracelsan interests led him into assumptions which were dangerous by the standards of the authorities.

Helmont never tired of criticising the orthodoxy of the universities and throughout his life he kept up a bitter attack on academic Galenic medicine. He rejected the Galenic doctrine of the humours as nonsense:

I have shown the vanity and falsehood of the device of Humours, whereby Physicians from a destructive foundation have circumvented the whole world, ... have destroyed families, have made widows and orphans by many ten thousands. (*Opera Omnia* p. 1013)

Not the least reason why Helmont rejected the teaching of the schools was its paganism, He did not think God had revealed the gift of healing to pagan authors. Hence whoever assented to the doctrine of 'Paganish schools' was excluded from 'the true principles of Healing'.

The medical interests of Helmont emerge most clearly in his treatises on the stone and on blood-letting. In both cases he rejected the beliefs on which Galenic remedies were based. But his own cures rested upon assumptions about God, Nature and Man. He stressed the overwhelming significance of the spiritual source of life:

Life is a Light and a formal beginning whereby a thing acts what it is commanded to act: but this light is given by the creator ... even as Fire is struck out of a Flint. (*Opera Omnia*, p. 744)

The soul leaves its stamp upon the flux of material existence.

The Fall of Adam, however, had changed the character of man's relationship to the natural world. Before the Fall, man's spiritual principle, his immortal mind, acted directly upon nature. After the Fall, the immortal mind was weakened and it was forced to act indirectly through the soul. The soul, in Helmont's view, ranked below the mind in significance. The task of the philosopher was to recognise the fundamental structure of reality and to create a situation in which the mind might operate as powerfully as possible upon the soul and ultimately upon the living principles or seeds (Archeus) of things. In Helmont's own words:

The mortal sensitive soul coexists in man (in his present state) with the immortal mind (*mens*), the soul being as it were the husk or shell of the mind, and the latter works through it, so that at the bidding of the mind, the soul makes use of the Archeus, whether it itself will or not. Before the Fall of Adam, man had only the immortal mind which acted directly on the Archeus, discharging all the functions of life, and man was immortal, the shadows of the brute beast not blurring his intellect. At the Fall, God introduced into man the sensitive soul and with it death, the immortal mind returning into the sensitive soul and becoming as it were its kernel.[30]

His assumptions about particular ferments and seeds in different organs of the body led him to assume different chemical processes and to foreshadow the modern concept of the enzyme. Helmont also seems to have been the first to describe the role of acid digestion in the stomach.

His anti-Galenic assumptions led him on to examine gases. He spoke of 'a spiritual gas containing in it a ferment . . . one thing is not changed into another without a ferment and a seed'. He described how the addition of nitric acid to sal ammoniac created a wild spirit which would burst a closed vessel. This he compared to the Greek 'chaos' from which he took the name gas. This particular gas, he called wild gas (*gas sylvestre*). He also distinguished other gases, all unknown to the Galenists, although some differed merely in name. They included *gas carbonum, gas pingue* and *gas sulfuris*. In all this scientific activity, Helmont asked questions which did not arise within the Aristotelian or mechanist traditions.

Helmont was conscious of belonging to the same tradition as Paracelsus though he rejected the Paracelsan concepts of 'macrocosm' and 'microcosm'. He referred to Paracelsus on many occasions as the first man to attack the Galenic doctrine of the humours. Helmont was also influenced by William

Gilbert. He quoted Tauler, the fifteenth-century religious writer, and his attacks on atheism won him the approval of the Cambridge Platonists. Helmontian doctrines enjoyed a certain vogue in mid seventeenth-century England and Helmont's son Francis Mercury paid a protracted visit there.

One of the attractions of Helmont's teaching at that moment of time was that it offered an answer to mechanism. Helmont's emphasis upon the sovereignty in nature of the immortal mind made an understandable appeal to those who felt that Christian teaching was in peril. God for the Helmontian was not an engineer but an artist who

As a painter, doth first conceive in his mind a spiritual Idea of the picture he intendeth to draw and afterwards by peculiar motions of his hand, which are guided by the said Idea, he produceth a Perfect picture corresponding with that in his mind.[31]

This quotation from an English Helmontian, Thomas Shirley, shows a typical emphasis upon the directing role of the mind.

We may conclude with a poem addressed by a contemporary to Helmont:

Shut up thy schools, O Galen, for enough of men are slain
So now it is sufficient: full graves do ring again
For Blood and Clyster are thy medicines: nothing oftentimes
Thou giv'st: but to a Critick day, thy hope alone confines.
In touching of a vein, the while, and eke of parched tongue,
And in the Urine wholly th'art dismayed, and so in dung.
A medicine's to be got for him, this helps not the sick man:
No need of tests of the Disease; but of a Physician,
Yet thou expect'st a great reward, after the man's enshrined
So doth the Dog look for and love, the cattle sickly kind.
Helmont is one, who able is by his Apollo's art
To snatch from th' jaws of Death whom t'other left to dye in smart.
(Helmont, *Opera Omnia*, p. 811)

Tycho Brahe's famous observatory, Uraniborg, on the island of Hveen off the coast of Denmark. Tycho (1546–1601) drew up elaborate horoscopes for members of the Danish royal family. At bottom left is the Emperor Ferdinand's villa at Prague, from which Tycho made observations during 1600-1.

Kepler

In the work of the German astronomer, Johannes Kepler (1571-1630), the magical tradition reached one of its climaxes. Kepler was a convinced Copernican in his youth, which marked him out as exceptional among contemporary astronomers and made him conscious of being a member of a small minority under pressure. It was for this reason that he urged Galileo to speak out openly in defence of the Copernican cosmology. The fact that Copernicanism was still very much an eccentric view in 1600 is worth stressing once more, for it establishes the important point that Kepler himself was an eccentric.

Kepler's role within the magical tradition was made distinctive by two factors. The first of these was the mass of astronomical observations which the Danish astronomer, Tycho Brahe (1546-1601), assembled and which Kepler was able to put to good use. Secondly, Gilbert's work on magnetism had a profound influence upon Kepler's formulation of his cosmological hypotheses.

Tycho's interest in astronomy sprang from a consuming obsession with astrology, which spilled over into natural magic. He devoted himself to constructing instruments which enabled him to re-map the night sky with an accuracy hitherto unknown, but this was not a detached scientific curiosity. It was an enthusiasm which rested on the assumption that more accurate knowledge of the stars and planets would lead to more accurate horoscopes. The information thus gathered he did not propose to share with his fellow 'scientists'. Tycho's observations were a private collection of data which he proposed to use to build up an unchallenged position of privilege in the world of astrology. On the basis of his observations, he succeeded in becoming the astrologer *par excellence*.

A diagram from Kepler's *Astronomia Nova* (1609) in which he used the observations taken of Mars by Tycho Brahe to argue in favour of Copernican cosmology. The three sketches illustrate the path of Mars according to Copernicus, Ptolemy and Tycho. Copernicus' heliocentric theory made the explanation simpler but his belief in circular orbits posed many false problems. Kepler himself proved the existence of elliptical orbits.

The career of Tycho is enormously interesting because it demonstrates how much of sixteenth-century interest in astronomy sprang from a quasi-religious belief in the importance of the stars. Tycho resembled the great scientists of the nineteenth century in his devotion to observation, but there the resemblance ceases. He was a mystic seeking his salvation in the night sky, and jealously guarding the results which he achieved. Kepler managed to make use of Tycho's data for his own purposes only by accepting the post of 'research assistant' in Tycho's entourage. The information was not freely volunteered for the sake of the advancement of learning and the Scientific Revolution. Tycho's science was essentially esoteric, in the great magical tradition.

If Tycho's observations provided Kepler with a means of testing his hypotheses, Gilbert's theory of magnetism provided him with a stimulus to his imagination. Following Gilbert, Kepler saw the earth as a great magnet, but he went further than this and applied the concept of magnetic attraction to the planetary system as a whole. Magnetism, as Kepler saw it, did away with the need for stellar intelligences among the planets in their orbits. It provided a hint that the mysterious links between sun and planets might be less powerful as the distance grew. In turning to Gilbert, Kepler was turning to a kindred spirit who provided evidence for Copernican hypothesis within the magical framework.

Kepler was born a Lutheran and was educated at the Lutheran university of Tübingen. As a poor scholar, his obvious choice of career was the ministry, but his religious beliefs were unorthodox, or thought to be by the authorities. He became a mathematics teacher in the Protestant seminary at Graz, which had ties with the university of Tübingen. While he occupied this humble teaching post (1594-1600),

Jam poſtquam ſemel hujus rei periculum fecimus, audacia ſubvecti porro liberiores eſſe in hoc campo incipiemus. Nam conquiram tria vel quotcunque loca viſa MARTIS, Planeta ſemper eodem eccentrici loco verſante: & ex iis lege triangulorum inquiram totidem punctorum epicycli vel orbis annui diſtantias a puncto æqualitatis motus. Ac cum ex tribus punctis circulus deſcribatur, ex trinis igitur hujusmodi obſervationibus ſitum circuli, ejúſque augium, quod prius ex præſuppoſito uſurpaveram, & eccentricitatem a puncto æqualitatis inquiram. Quod ſi quarta obſervatio accedet, ea erit loco probationis.

PRIMVM tempus eſto anno MDXCX D. v Martii veſperi H. VII M. x eo quod tunc ♂ latitudine penè caruit, ne quis impertinenti ſuſpicione ob hujus implicationem in percipienda demonſtratione impediatur. Reſpondent momenta hæc, quibus ♂ ad idem fixarum punctum, redit: A. MDXCII D. XXI Jan. H. VI M. XLI: A. MDXCIII D. VIII Dec. H. VI. M. XII: A. MDXCV D. XXVI Octob. H. v M. XLIV. Eſtq; longitudo Martis primo tempore ex TYCHONIS reſtitutione i. 4. 38. 50: ſequentibus temporib. toties per i. 36 auctior. Hic enim eſt motus præceſſionis congruens tempori periodico unius reſtitutionis MARTIS Cumq; TYCHO apogæum ponat in 23 ½ ♌, æquatio ejus erit 11. 14. 55: propterea lógitudo coæquata anno MDXC i. 15. 53. 45.

Eodem verò tempore & commutatio ſeu differentia medii motus SOLIS a medio Martis colligitur 10. 18. 19. 56: coæquata ſeu differentia inter medium SOLIS & MARTIS coæquatum eccentricum 10. 7. 5. 1.

PRIMVM hæc in forma COPERNICANA ut ſimpliciori ad ſenſum proponemus.

Sit a punctum æqualitatis circuitus terra, qui putetur eſſe circulus δ γ ex a deſcriptus: & ſit Sol in partes β, ut a β linea apogæi

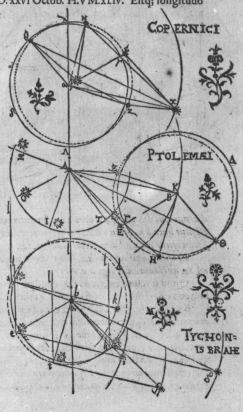

COPERNICI

PTOLEMAEI

TYCHONIS BRAHE

Kepler wrote his first major book, *Mysterium Cosmographicum*. He wrote it before being influenced by Gilbert and before meeting Tycho. Though it was much less of a piece of cosmological speculation than the title suggests, Kepler claimed to show in it that the spheres of the six planets (including the earth) in the Copernican system corresponded to the five perfect solids of Euclid: four-sided, six-sided, eight-sided, twelve-sided and twenty-sided figures:

The earth is the measure for all other orbits. Circumscribe a twelve-sided regular solid about it; the sphere stretched about this will be that of Mars. Let the orbit of Mars be circumscribed by a four-sided solid. The sphere which is described about this will be that of Jupiter. Let Jupiter's orbit be circumscribed by a cube. The sphere described about this will be that of Saturn. Now place a twenty-sided figure in the orbit of the earth. The sphere inscribed in this will be that of Venus. In Venus' orbit place an octahedron. The sphere inscribed in this will be that of Mercury. There you have the basis for the number of planets.[32]

The interest of this passage is twofold. It reveals Kepler the Pythagorean attempting to apply to the cosmos the kind of mathematical insights which Pythagoras had discovered in music. It also shows us Kepler the Copernican accepting the earth as a planet. The two roles cannot be distinguished. Kepler was not trying merely to describe the paths of the planets. He was attempting to explain *why* there are only six planets. In other words, he was providing an insight into God's mind. God created the cosmos upon the basis of the divinely inspired laws of geometry. Hence Kepler's conclusions were mystical and geometrical at one and the same time. It is true that the *Mysterium* is early Kepler, written by a young man, but the same tone is to be found in his later writings.

Kepler became convinced in 1595 that the structure of the planetary system was based upon the five regular polyhedra. He tried to persuade Frederick Duke of Württemberg to have a drinking cup made after the model of the universe. Each planetary sphere would provide its own beverage, ranging from the sun's *aqua vitae* to Saturn's 'bad old wine'. This plan was dropped in favour first of a globe, then of a mobile planetarium. Nothing came of Kepler's enthusiasm.

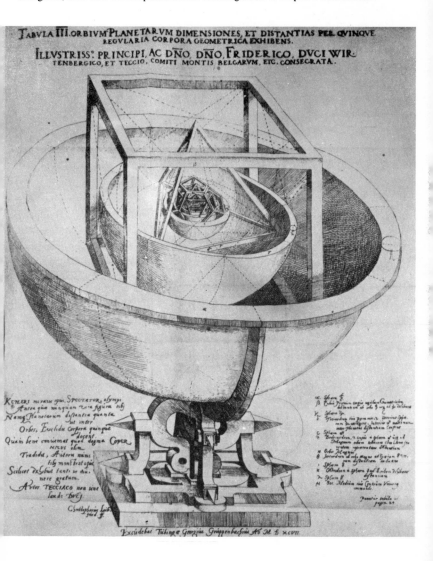

Portrait of Johannes Kepler (1571-1630), whose astronomical discoveries provided a vital transition between the world of Tycho Brahe and the world of Newton.

Kepler's next book *Astronomia Nova* was written around 1605-6, though not published until 1609. By then he had moved to Prague first in the service of Tycho and then as Imperial Astronomer to Rudolph II. It was during this period that he read Gilbert's *De Magnete* and was strongly influenced by it. It seems also that after writing the *Mysterium* Kepler devoted more time to mathematics.

The *Astronomia Nova* was the first revolutionary piece of astronomy since Copernicus's *De Revolutionibus*. In it, Kepler put forward two radical innovations. First he argued that the planets moved in elliptical orbits round the sun, and secondly, that the velocity of the planets was not uniform in the course of their individual orbits. These were the bases of Kepler's first two laws, and they owed a great deal to the influence of Tycho and Gilbert. Tycho's data enabled him to test his hypotheses and to reject them when they failed to fit the facts. It was indeed only after many formulations that Kepler was driven to accept the ellipse as the planetary orbit. Gilbert's magnetic theories provided him with an idea which enabled him to accept the possibility of the planets moving at varying speeds, which could be explained in terms of the weakening magnetic interraction at greater distances between them and the sun.

In addition to his debt to Tycho and Gilbert, Kepler owed an immense amount to the sustaining power of his Pythagorean belief in the mathematical harmony of the universe. Kepler refused to give up when failure followed upon failure and the facts did not fit his successive hypotheses. This persistence had its origins in his quasi-religious belief that God had created the universe in accordance with the laws of mathematics.

Kepler's first two laws radically modified not only the Ptolemaic system but also Copernicanism, as originally set out in 1543. Copernicus had retained the notion of circular motion which inevitably involved holding on to epicycles as part of his explanation. He also assumed that the planets moved at uniform speed. Kepler destroyed both of these assumptions, or he destroyed them in theory. In fact his *Astronomia Nova* had remarkably little influence. Kepler was too neo-Platonic for most of his fellow astronomers and he was too mathematical for his fellow neo-Platonists. This book which appears so revolutionary to us was perhaps not appreciated until Sir Isaac Newton saw its true value over fifty years after its publication. Galileo and Descartes, the most influential of all seventeenth-century scientists before Newton, accepted as axiomatic that the planets moved in circular orbits at uniform speed and hence dismissed Kepler's theories as unfounded speculation.

Kepler's third major book, *Harmonice Mundi*, published in

1619, was the fruit of his years spent as mathematician in the Protestant school at Linz, where he moved in 1612 after the death of Rudolph ii. In this book, Kepler, building upon earlier work but perhaps also in reaction to the restricted circumstances of his daily existence, attempted a neo-Platonic synthesis, and sought to reveal the mathematical language of the creator in almost every aspect of the universe. The interest of the book to later astronomers lay in the exposition of Kepler's third law, that there is a constant ratio between the square of the planetary period of revolution and the cube of the planets' mean distance from the sun. But this equation, proving that God was a mathematician after all, was almost hidden in a mass of neo-Platonic speculation.

What the historian cannot do is to separate Kepler the 'scientist' from Kepler the neo-Platonic mystic. Kepler would not have been moved to question the basis of existing cosmological theories unless he had been a neo-Platonist in the beginning. The *Mysterium* shows us this. But the tone of the *Mysterium* of 1597 is to be found in Kepler's language twenty years later when he described how

I feel carried away and possessed by an unutterable rapture over the divine spectacle of the heavenly harmony.[33]

He spoke in *Harmonice Mundi* (1619) of how 'the configurations strike up: sublunary nature dances after the fashion of music'.[34]

In *Harmonice Mundi*, Kepler applied the analogy of modern polyphonic music to the planets. Each planet had its own scale which was determined by its speed and the musical climax was a chord sounded by all six planets. Thus Kepler's God was an artist who took satisfaction in the musical instrument he had created. 'The motions of the heavens', he wrote, 'therefore, are nothing else but a perennial concert, made up of rational [unheard] music'. If polyphonic music gave such pleasure the reason was because it was a reflection of celestial music.

Kepler, like Gilbert and Bruno, believed that the sun was a soul. His early Copernicanism revealed in the *Mysterium* rests upon the idea that the sun as a spiritual 'force' acting at a distance was responsible for moving the planets in their orbits:

If we want to get closer to the truth and establish some correspondence in the proportions (between the distances and velocities of the planets) then we must choose between these two assumptions: either the souls which move the planets are less active the further the planet is removed from the sun, or there exists only one moving soul in the centre of all the orbits, that is the sun, which drives the planets the more vigorously the closer the planet is, but whose force is quasi-exhausted when acting on the outer planets because of the long distance and the weakening of the force which it entails.[35]

He believed that there was an earth-soul which perceived geometrical relationships and which expelled its 'subterranean humours' (our volcanoes and earthquakes of today), when planetary rays met at an appropriate angle:

It is the custom of some physicians to cure their patients by pleasing music. How can music work in the body of a person? Namely in such a way that the soul of the person, just as some animals do also, understands the harmony, is happy about it, is refreshed and becomes accordingly stronger in its body. Similarly the earth is affected by a harmony and quiet music. Therefore there is in the earth not only dumb, unintelligent humidity, but also an intelligent soul which begins to dance when the aspects pipe for it. If strong aspects last, it carries in its function more violently by pushing the vapours upwards, and thus causes all sorts of thunderstorms; while otherwise when no aspects are present, it is still and develops no more exhaltation than is necessary for the rivers.[36]

Kepler saw the role of the scientist as akin to that of the priest or the seer; the poet, the lover and the scientist were of imagination all compact. He related in *Harmonice Mundi* how

I gave myself up to sacred frenzy. I have plundered the golden vessels of the Egyptians, in order to furnish a sacred tabernacle for my God out of them far from the borders of Egypt.[37]

Kepler saw God not as the logician or the engineer but as the playful magician leaving his marks in the universe for us to discover. The world of nature carried signs, or signatures, left by God as clues to indicate their true significance, or utility. Thus a plant designed to cure a particular disease, would carry a 'signature' indicating this:

God himself was too kind to remain idle and began to play the game of signatures signing his likeness on to the world; therefore I chance to think that all nature and the graceful sky are symbolised in the art of Geometria.[38]

In Kepler's scientific approach we recognise a distinctive style which has as much in common with Pico, Lull, and Paracelsus as with Galileo. He resembled Michelangelo more than Leonardo in his search for the unseen harmony of nature. It is true that he clashed with the Englishman Robert Fludd over Fludd's Hermetic and mystical use of numbers. But Kepler and Fludd from our vantage point seem to be far closer than they appeared to themselves or their contemporaries. Kepler was a mathematician of genius but he thought 'more Hermetico', after the manner of the neo-Platonists.

5 The world as machine

Galileo

The mechanistic interpretation of natural phenomena, as we have seen (page 41), had its origins in Renaissance Italy. The machines of Leonardo, the mechanical interests of Nicholas Tartaglia and the revival of Archimedes appear as part of the same complex of ideas, which stressed the predictable interaction of mechanical forces in nature. Leonardo, for example, in his flying machine attempted to reproduce in mechanical terms the natural flight of birds. But during the sixteenth century the mechanist approach was confined to a minute area of experience. It was not until the next century, beginning with Galileo and then proceeding further with Mersenne and Descartes, that the idea of seeing the *whole* of nature in mechanistic terms took hold.

Modern mechanism began with Galileo, but he did not invent the notion. One of the decisive influences upon his scientific outlook was undoubtedly Archimedes, whom he mentions over a hundred times, often in terms of the utmost reverence ('the most divine Archimedes'). Galileo was also an intellectual disciple of Tartaglia, though he applied Tartaglia's methods to a wider range of theoretical problems. This mechanistic approach made itself evident in Galileo's early student days at the university of Pisa and was still dominant in his *Discourses concerning the Two New Sciences* published in 1638, towards the end of his life.

Galileo is largely remembered in the textbooks for his struggle with the Papal Inquisition over the Copernican theory of the universe. This episode has all the qualities of great drama, but it perhaps is misleading as a guide to the working of Galileo's mind. It tempts us to place him along with Copernicus, Bruno and Kepler within the neo-Platonic tradition of

Galileo (1564-1642) presenting his telescope to the muses and pointing out a heliocentric system. The implications are that astronomy was still one of the liberal arts.

natural philosophy, when the truth seems to be that Galileo appropriated the heliocentric cosmology for his own purposes and placed it within an entirely different frame of reference. What Descartes was to do later with Harvey's circulation of the blood, Galileo did on a grander scale. In a sense, he caught the Copernicans bathing and ran away with their clothes.

In astronomy, Galileo's chief claim to fame was to see the implications of an optical instrument which had been invented by someone else. Galileo had the imagination to use the telescope to look at the planets and to announce his discoveries to a wide audience in a short pamphlet *Message from the Stars* (1610). The telescope enabled him to discover the moons of Jupiter and from this he concluded that he had proof of the Copernican hypothesis. Galileo had none of the neo-Platonists' worship of the sun or their belief in the earth soul. He was led to his acceptance of Copernicanism by a mechanical analogy drawn from his astronomical observations. He did not accept Kepler's theory of elliptical paths for the planets or the non-mechanical notion of magnetic attraction at a distance. The secret of planetary motion, as Galileo saw it, was the tendency of bodies to follow a circular path. 'Hence I think it may be rationally concluded that, for the maintenance of perfect order among the parts of the universe, it is necessary to say that bodies are moveable only circularly.'

Galileo did not take the neo-Platonists as his main target. He saved the weight of his sarcasm for the Aristotelians. But his attitude towards the assumptions of Kepler and Bruno emerges clearly enough in passages like the following, dealing with tides:

I cannot be brought to subscribe to lights, to temperate heats, to predominances by occult qualities and to such like vain imaginations

Galileo maintained that God was a
mathematician who constructed the
universe as a mathematical model.
Hence the key language was mathematics,
of which this page, taken from
Galileo's *Dialogues*, is an example.

that are so far from being, or being possibly, causes of the tide, that, on the contrary the tide is the cause of them.[39]

Galileo was also critical of Gilbert and even though he was impressed by his discoveries in magnetism, he rejected the neo-Platonic interpretation which Gilbert placed upon them:

That which I could have desired in Gilbert is that he had been a somewhat better mathematician and particularly well grounded in geometry, the practice whereof would have rendered him less prone to accepting for true demonstrations the reasons he produces as causes of the true conclusions observed by himself.[40]

Thus we may describe Galileo as poised between two distinct traditions, the Aristotelian and the neo-Platonic, not merely engaged, as we are often told, in single combat with the Aristotelian world.

It is tempting to believe that Galileo also took the Copernican theory out of its quasi-religious framework and explained it in secular, mechanical terms. But this is true only up to a point. Galileo, like all seventeenth-century scientists, had an idea of God which was fundamental to his interpretation of the universe. He saw nature as reflecting the mind of the Deity, and since Galileo's natural world was mechanist, Galileo's God was inevitably a craftsman, albeit a Divine one. He refers to the art and skill of God:

The turning to the great volume of Nature, which is the proper object of philosophy . . . in which book, . . . being the work of Almighty God . . . is more absolute and noble wherein most greatly is revealed his art and skill.[41]

He speaks of God as the Divine Architect arranging the cosmos as seemed best:

Let us suppose that among the purposes of the Divine Architect is the creation of these continually moving globes and the assign-

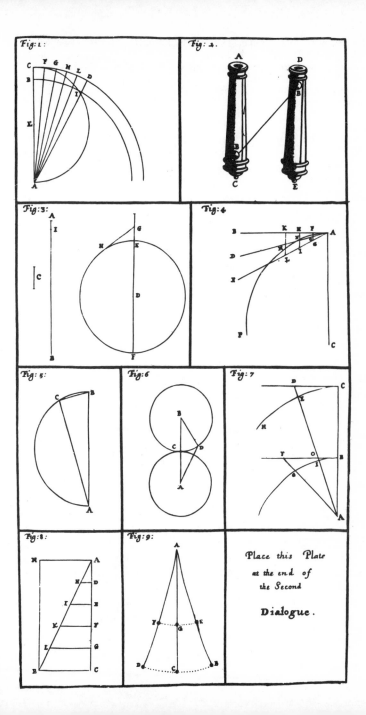

Place this Plate
at the end of
the Second

Dialogue.

The first map of the moon made in 1645 by Johan Hevel (or Hevelius), whose *Selenographia* 'Drawing of the Moon' has been described as 'among the finest products of seventeenth-century astronomy'. The map shows the dark areas, which had been known as 'seas' since Galileo's time (e.g. the Sea of Storms, top left) and the craters which were given names in about 1650 (e.g. Tycho, bottom centre and Copernicus, left centre).

ment of the immovable sun as the centre of their revolutions; we may assume that He made all the globes in the same place and with the intended inclinations to move towards the centre till they had acquired those velocities which at first seemed good to the Divine Mind.[42]

His God was also a mathematician:

I say that human wisdom understands some propositions as perfectly and is as absolutely certain thereof, as Nature herself; and such are the pure mathematical sciences, to wit Geometry and Arithmetic. In these the Divine Wisdom knows infinitely more propositions, because it knows them all . . .[43]

The way in which Galileo plays 'the divine card' here is breathtaking, but from our point of view its significance is an insight into Galileo's own mind. Like the neo-Platonists and the Aristotelians, the mechanists had their own concept of the Divine nature which they used to support their theories. Galileo used the next world to redress the balance of his own.

Galileo's mechanistic interpretation of nature placed him on a collision course with the Aristotelians, even if he had not ventured into astronomy, since it implied a rejection of the Aristotelian emphasis on final causes. In his *Dialogues* he criticised 'final causes' by implication by placing the defence of the concept in the hands of a ridiculous character Simplicio, whose simplicity is continually exposed. Simplicio's defence of final causes ran as follows:

[Final causes exist] Because we clearly see that all generations, corruptions, etc. made in the Earth are all either mediately or immediately directed to the use, convenience and benefit of man; horses are brought forth for the use of man, the earth produces grass for the feeding of horses, and the clouds water it . . . the end to which [all things are] directed [are] the necessity, use, convenience and benefit of man.[44]

He also used the telescope to expose the unreality of the Aristotelian belief that the heavens were perfect. As he makes Simplicio say:

The celestial bodies being eternal, inalterable, impassible, immortal etc they must necessarily be perfect.[45]

In all these crucial instances, Galileo attacked the Aristotelian position. Where he left himself open to counter attack was in stating openly the implications of his position for orthodox interpretation of the Scripture.

Galileo also rejected the Aristotelian qualitative division of motion into 'natural' and 'unnatural' motion. He was able to prove, in *Two New Sciences*, that the motion of a projectile was not a simple form of motion, but followed a path dictated by a quasi-mechanical interaction of two different kinds of motion. Galileo's experiments took him out of the 'real' Aristotelian world of qualities into an abstract world in which quantitative differences were the primary factor.

Such criticisms had implications far beyond their scientific interest, since Aristotelianism was not merely an academic movement but an entrenched and powerful orthodoxy, especially in Italy. Within the counter-reformation church, the Council of Trent (1545-63) tightened up ecclesiastical discipline and clarified doctrine, as an answer to the challenge of Luther and Calvin. A decisive role in all this was played by the Jesuit order, founded in 1540, which adopted Aristotle and Aquinas as its official guides in philosophy and theology. The Jesuits dominated the theological discussions at Trent, with the result that Aristotelian philosophical concepts were used in official theological definitions. In the half century following Trent, the Jesuits, backed by the Papacy, were the educational instrument behind an Aristotelian revival. At the same time, the Holy Office, founded by the Papacy in 1542, acted as the watchdog of orthodoxy and of Aristotelianism. With its headquarters in Rome, the Holy Office controlled most of the intellectual life of Italy. The exception was the Republic of Venice, where a more liberal atmosphere survived. Anyone thought to hold dangerous opinions was liable to be brought before the Holy Office for trial and punishment, possibly death, as Bruno discovered in 1600.

Thus in criticising the Aristotelians, Galileo was not attacking the enfeebled remnants of an outworn philosophy. On the contrary, he took an immense risk. His *Dialogues* published in 1632 were not technical astronomical treatises like Copernicus's *De Revolutionibus*, they were a piece of brilliant polemic, aimed at the clerical 'establishment'.

Galileo counted on the support of his patrons, the Barberini Pope Urban VIII and the Archduke of Tuscany. In both cases his confidence proved mistaken. Urban was a broken reed in the face of pressure by the Inquisition, and in Tuscany, the Archduke died and Galileo, left exposed to his enemies, was forced to capitulate. If we wish, we may see this primarily as a victory of intolerance over intellectual freedom. More significantly, from our point of view, the clash was between two paradigms, the organic and the mechanistic. The Aristotelians were successful in 1633, but before the end of the century, they were in a state of disarray.

The tradition of Galileo survived in Italy for one more intellectual generation. In the work of Evangelista Torricelli (1608-47) Galileo's mechanistic approach was carried into the study of air pressure. The crux of the Torricellian experiment, using a column of mercury, was to demonstrate that air, one of Aristotle's four elements, behaved according to mechanical laws. Torricelli thus extended the analogy of a machine into another field of natural investigation. The radical implications of his work were seen both by Mersenne and Pascal (see page 163). Torricelli's simple experiment (see page 74) raised the question of the possibility of a vacuum in nature, which on Aristotelian premises (as well as Cartesian) was a contradiction in terms.

After Torricelli, the mechanist tradition was carried on by his successors at Florence, Viviani (1622-1703) and Alfonso Borelli (1608-79). Of these the most notable was Borelli. He carried mechanical interpretations into the fields of anatomy and astronomy and developed a theory of planetary movement which was based on the assumption that the planets were driven round by light rays emanating from the sun.

Mersenne

In the spread of mechanistic ideas outside Italy, the crucial link was an unlikely figure – the French friar, Marin de Mersenne (1588-1648). Mersenne was an enthusiastic admirer of the mechanistic outlook of Galileo. In 1634 he published a French translation of Galileo's early lectures (1592) on mechanics, and in 1639, a year after their original publication, he translated Galileo's *Discourses on Two New Sciences*. But he was not a Copernican and he took over Galileo's mechanical assumptions without his cosmology. He did *not* translate Galileo's *Dialogue on the Two World Systems* when it appeared originally in 1632, presumably because of its condemnation by the Holy Office, though he summarised parts of it.

Mersenne was no sceptic. He was horrified by the thought that there were many thousands of unbelievers in the Paris area and he saw the mechanical philosophy as a way of refuting their unbelief. In this, he resembled Descartes, whose own dissatisfaction with scholasticism stemmed from its inability to meet the sceptics on their own ground.

Mersenne became the central figure in a network of correspondents which stretched over France, Holland, Italy, England and the Spanish Netherlands and which provided a focal point for an informal discussion on natural philosophy. In this way he acted as a channel for the spread of mechanistic ideas, though he himself produced no scientific masterpiece.

Part of the interest of Mersenne lies in the full-blooded attack which he launched upon the magical tradition. In this, he provides a contrast with Galileo. Mersenne reserved his main ammunition for the neo-Platonists, while tacitly ignoring the Aristotelians, whereas Galileo did the opposite. Mersenne rejected the occultism of Bruno and Campanella (1563-1639). He actually met Campanella in Paris in 1634 and described him as having 'a happy memory and a lively imagination', but with nothing to offer in the way of science. Mersenne was also very critical of his English contemporary Robert Fludd, the most prolific exponent of neo-Platonic ideas in the 1630s.

Another link between Mersenne and Galilean mechanism was provided by Torricelli. During a visit to Italy in 1645, Mersenne met Torricelli and discussed with him his experiments in connection with the vacuum. Mersenne returned to France and conceived the idea of pushing the experiment further by taking the tube of mercury up a mountain and observing the result. He wrote to a friend at Puy de Dôme asking him to arrange for this to be done, but he eventually learned that his friend had moved to another part of France. Shortly afterwards Pascal's famous vacuum experiment took place on the same mountain and Mersenne missed the chance of scientific fame. In the light of this there is inevitably a sense of anti-climax about his career, but in recent years his significance as a catalyst in making intellectual change possible has come to be recognised.

Descartes and the spread of mechanism

Of all the mechanists of the seventeenth century the most influential was René Descartes (1596-1650) for the simple reason that his *Discourse on Method* (1637) was a short treatise, written in an autobiographical form, with a clearly expressed thesis. The *Discourse* combined literary grace, human interest and philosophical clarity in a form not seen in Europe since Plato's *Dialogues*, and was unrivalled for popularity save by Galileo's *Message from the Stars*. Descartes appealed to a new phenomenon, the philosophically minded gentlemen, doctors and lawyers, whose status rose during the century at the expense of the clergy. His fellow mechanist Mersenne was still part of a clerical tradition stretching back to the middle ages. Descartes, in contrast, looked ahead to the 'philosophes' of the eighteenth century.

Despite this, Descartes had much in common with Mersenne. His *Discourse on Method*, like all his later writings, had a religious as well as a philosophical end in view. Descartes may have been the father of scepticism, but it is a wise father that recognises his own child in later years. What Descartes was aiming at was a refutation of scepticism and his four points of philosophical method were intended to clear the ground on which a more substantial structure than scholasticism could be erected. His famous tag 'Cogito ergo sum', ('I think, therefore I am'), was a religious affirmation. Thought was the activity proper to the soul. It was a spiritual activity in a mechanical universe and hence an answer to those sceptics who refused to accept the existence of the soul. In fact, we do no injustice to Descartes if we see him as a second Aquinas creating a synthesis of all knowledge and bringing Christian Revelation and new learning in a fresh whole.

Descartes' concept of God emphasised power and truth rather than love and goodness. He assumed that the deity resembled an ingenious engineer, on Archimedean lines, an idea which the following passage from his *Principia Philosophiae* brings out:

Just as the same artisan can make two clocks, which though they both equally well indicate the time, and are not different in outward appearance, have nevertheless nothing resembling in the composition of their wheels; so doubtless the Supreme Maker of things has an infinity of diverse means at his disposal, by each of which he could have made all the things of this world to appear as we see them, without it being possible for the human mind to know which of all these means he chose to employ.[46]

The concept of a Divine Engineer is also implied in Descartes' use of mechanical analogies to describe what God has created. He tells us that 'the rules of mechanics . . . are the same with those of nature', and in discussing the proof of God's existence, he uses the analogy of a machine-maker:

There is no difference between this and the case of a person who has the idea of a machine, in the construction of which great skill is displayed, in which circumstances we have a right to inquire how he came by this idea, whether for example, he somewhere saw such a machine constructed by another, or whether he was so accurately taught the mechanical sciences, or is endowed with such force of genius that he was able of himself to invent it, without having elsewhere seen anything like it; for all the *ingenuity* which is contained in the idea objectively only, or as it were in a picture, must exist at least in its first and chief cause.[47]

In the history of philosophy, Descartes has his place as the first modern critical thinker. In the history of science, his significance lies in the fact that he was the first to construct a scientific system, which conflicted at almost every point with Aristotelian principles. Descartes rejected the basic assumption that the laws of the lunar world were different from those of the sub-lunar world. In his view, the stars, planets and earth were made of the same substance ('it is one and the same matter which exists throughout the universe'). He rejected the Aristotelian principle that natural motion was directed towards an end, and hence purposive (final causes). For Descartes, motion was not the movement from place to place within a universe that had an absolute 'up' and 'down'; it was merely 'the translation of a piece of matter from the

Cartesian science is now almost forgotten
but the importance of Descartes (1596-1650)
in the overthrow of Aristotelianism
and the substitution of mechanism
can scarcely be over-estimated.

neighbourhood of the bodies touching it, to the neighbour-hood of others'. This cut away the ground from Aristotle's distinction between 'natural' and 'unnatural' motion. Descartes explained the movement of planets in terms of the swirling movement of elliptical vortices of matter, thus attacking Aristotelian assumptions about the primacy of circular motion and the quintessential nature of planetary substance. For the Aristotelian theory of the four elements, he substituted a theory of particles. On this basis he explained chemical change in terms of a mechanical adjustment of particles. This implied rejecting Aristotelian emphasis on qualities and the persistence of substantial forms.

One of the key features of the Cartesian universe was motion. Descartes took this as axiomatic though he was only able to explain it in terms of divine action:

He created matter along with motion and rest in the beginning; and now, merely by his ordinary co-operation, he preserves just the quantity of motion and rest in the material world that he put there in the beginning.[48]

Since motion was a built-in feature of the universe, Descartes was free from the necessity of explaining it. By this means he escaped from Aristotelean pre-occupation with such questions as 'impetus'. On the other hand, he could not escape the influence of Aristotle completely. His very model for a synthesis was Aristotelian. He also made considerable use of deductive reasoning in constructing a picture of the universe. Indeed the empirical basis of his synthesis was as narrow as Aristotle's had been. Descartes drew back from the mechanistic implications of atomism and he did not come out fully in support of the Copernican heliocentric theory. In addition, he, as much as any Aristotelian, rejected the possibility of a vacuum.

The weaknesses of the Cartesian approach were to become apparent in the course of the century. His system of deductive reasoning left him as exposed to experimental attack as any of the scholastics. His doctrine that animals were machines failed to carry conviction. Most serious of all, in view of the fact that he was a mathematician of genius, his theories of planetary motion were not susceptible to mathematical demonstration. The combination of deductive reasoning and mechanistic assumptions may be seen in the following extract from the *Principia*, where Descartes asserts the existence of particles, even though they were not empirically observable:

My assigning definite shapes, sizes and motions to insensible particles of bodies, just as if I had seen them, and this in spite of admitting that they are insensible, may make some people ask how I can tell what they are like. My answer is this. Starting from the simplest and most familiar principles which our minds know by their innate constitution, I have considered in general the chief possible differences in size, shape, and position between bodies whose mere minuteness makes them insensible, and the sensible effects of their various interactions. When I have observed similar effects among sensible objects, I have assumed that they arose from similar interactions of insensible bodies; especially as this seemed the only possible way of explaining them. And I have been greatly helped by considering machines. The only difference I can see between machines and natural objects is that the workings of machines are mostly carried out by apparatus large enough to be readily perceptible by the senses (as is required to make their manufacture humanly possible), whereas natural processes almost always depend on parts so small that they utterly elude our senses. But mechanics, which is a part or species of physics, uses no concepts but belong also to physics; and it is just as 'natural' for a clock composed of such-and-such wheels to tell the time, as it is for a tree grown from such-and-such seed to produce a certain fruit. So, just as men with experience of machinery, when they know what a machine is for, and can see part of it, can readily form a conjecture about the way its unseen parts are fashioned; in the same way, starting from sensible effects and sensible parts of bodies, I have tried to investigate the insensible causes and particles underlying them.[49]

The Cartesian universe was mechanical in the sense that it existed as a machine and *nothing else*. Descartes stripped away from his view of the universe all that was extraneous to its mechanical functioning. It became the equivalent of a blue-

A sketch of the Cartesian universe
showing Descartes' conception of matter
swirling in vortices. This theory
enjoyed its heyday in the late seventeenth
and early eighteenth centuries, before
Newtonianism was finally accepted.

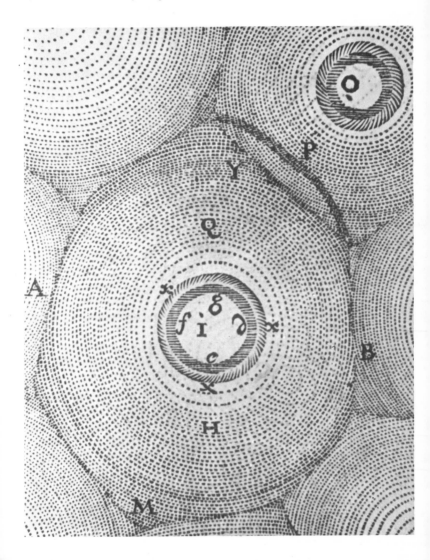

Parabolic path of a projectile demonstrated with a tennis ball in a letter from Descartes to Mersenne. This is an example of informal correspondence between scientists during the first half of the seventeenth century which later became formalised in philosophical societies.

print which was transformed into matter (extension). It was thus more mechanical than a machine, which at least possessed certain qualities such as colour. The Cartesian universe, therefore, was a machine stripped down to its absolute essentials. Descartes performed an act of abstraction on a cosmic scale, equivalent to that which Galileo had done with the ball rolling down the inclined plane. For Galileo, the ball and the plane were really irrelevant, as he was thinking beyond them to a mathematical universe. The same may be said of Descartes. The Cartesian universe was a mathematical one, organised on the basic principles of mechanics. Apparent qualitative differences within it were due to differences of motion.

This emphasis upon mechanical principles is to be found throughout Descartes' *Principia*. He explained light and heat, for example, in terms not of Aristotelian qualities but as the products of motion:

We must therefore conclude on all counts that the objective external realities that we designate by the words 'light', 'colour', 'odour', 'flavour' or 'sound', or by the name of tactile qualities such as 'heat' and 'cold', and even the so-called 'substantial forms' are not recognisably anything other than the powers that objects have to set our nerves in motion in various ways, according to their own varied disposition.[50]

In one respect, Descartes performed a magnificent sleight of hand. He transferred Harvey's discovery of the circulation of the blood, which had been set within the framework of Aristotelianism, and made it into the corner stone of his own view that the human body was a machine:

But lest those who are ignorant of the force of mathematical demonstrations and who are not accustomed to distinguish true reasons from mere verisimilitudes, should venture, without examination to deny what has been said, I wish it to be considered that the motion

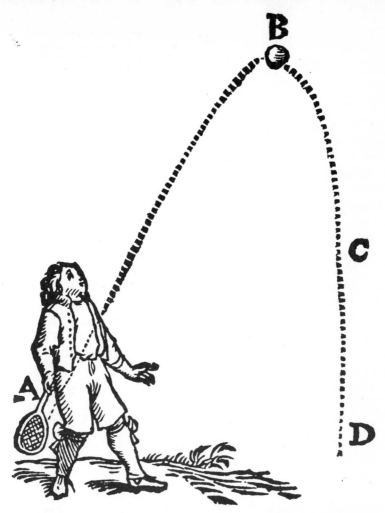

which I have now explained follows as necessarily from the very arrangement of the parts, which may be observed in the heart by the eye alone, and from the heat which may be felt with the fingers, and from the nature of the blood as learned from experience, *as does the motion of a clock from the power, the situation and shape of its counterweights and wheels.*[51]

As we have seen above, Harvey did not speak of the heart

An illustration from the first edition of Descartes' *Tractatus de Homine* (published posthumously in 1662 at Leyden) showing his views on sensation. The pineal gland is shown as the intermediary conveying the physical sensation of warmth to the mind.

in terms of a clock. He used images like 'sovereign' which supported his belief that the heart was the dominant organ of the body. For Descartes, however, the heart was no more important than any other part of the human machine, though it may be noted that even he did not think of the heart as a 'pump', but as a mechanism for heating the blood and so producing a kind of effervescence which led the blood to push itself out of the heart.

In his *Discourse on Method* Descartes made explicit the mechanistic assumptions that were merely implicit in Galileo and Mersenne. Between this view and the Aristotelian and neo-Platonic conceptions of the role of the natural philosophy there was an immense gulf. There was no mystery in the Cartesian world; there was no harmony of the spheres, there were no final causes. As Descartes wrote:

We will not seek reasons of natural things from the end which God or nature proposed to himself in their creation (i.e. final causes) for we ought not to presume that we are sharers in the counsels of the Deity, but [consider] him as the efficient cause of all things.[52]

The task of the natural philosopher was not that of the biologist or the seer, but simply to explain the working out of the mechanical principles on which the Divine Engineer had created the world.

For a century at least (1640-1740), Cartesianism was a powerful influence upon western European science. It was the most important single factor in the eventual victory of the mechanist tradition. It offered a rallying point for all the critics of the organic and magical traditions, it became the basis of a new scientific paradigm, which, though open to criticism in detail, offered great intellectual satisfaction. Above all, it was a victory for mathematical models of nature.

gitandum est, modum illum, quo aperiret tubum 7, in cau-
sa fore, ut partes cerebri, versus N comprimerentur, &

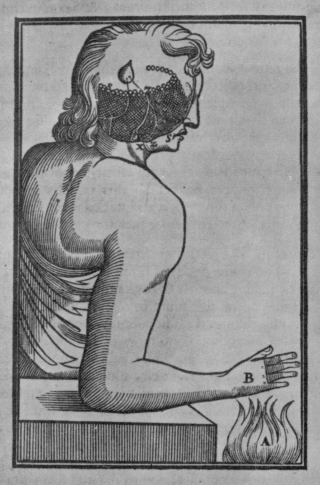

versus O dilatarentur paulo magis quam solent; atque ita
spiritus, qui veniunt à tubo 7 ab N per O versus P irent.
Posito autem quod hic ignis manum urat, actio ejus tubum

ᵃ *Ut nulla prorſus mutatio interveniat prӕterquam in ſitu glandulӕ H*] Hoc eſt, mutatio ſitus glandulӕ, ſive procedendo, ſive retrocedendo, erit in cauſa, quod anima poterit objecta diverſa diverſimode ſita ſentire, nec ulla obveniat mutatio organo exteriori, neque in modo, quo po-

rus 8 reſpicit glandulam, veluti in exemplo propoſito, progreſſio glandulӕ efficit, ut objecta N & O ſentiantur, quӕ ſecus nequiviſſent ideas ſuas diſtincte imprimere punctis *n* & *o* glandulӕ, organo exteriore & tubo 8 eo, quo ſunt, modo diſpoſitis.

The Cartesian conception of the relationship
between mind and body. Descartes argued that
the pineal gland, indicated here as H,
provided the crucial link between the body,
which operated on mechanical principles, and
the mind, which was non-material.

163

Pascal

The Mersenne circle included some of the most gifted men of the century, Hobbes and Descartes among them. Of them all, Blaise Pascal (1623-62) had perhaps the most universal genius. He was mathematician, experimenter, prose-writer and pamphleteer. He displayed his mathematical gifts at an early age, and he devoted a good deal of his time to inventing machines, including a calculating machine.

Pascal's place within the mechanistic tradition is reflected in his theology. He saw no trace of the Christian deity in the world of nature. His God was a hidden God, whose will was manifested not in the mechanical laws of nature but in miraculous intervention with these laws. His own religious conversion seemed to him one of these. So also was the miracle of the Holy Thorn which restored his niece to health in 1656. Pascal's God also intervened dramatically and decisively in his choice of the elect. The world of nature and of the spirit were thus in marked contrast. In the one, the world of matter followed mechanical laws, in the other, the unexplained and inexplicable will of God was the key.

In the Scientific Revolution, the interest of Pascal lies in the experiments which he recorded to draw conclusions about the behaviour of one of the four Aristotelian elements – air. According to the Aristotelians, air like fire was 'light', and therefore had no weight. This principle, as we have seen (see page 30) formed part of the Aristotelian world-picture and hence to call it into question was to challenge the whole system. But Pascal's account of an experiment with a balloon serves as an example of the simplicity and imaginativeness of his approach.

Pascal invented his calculating machine when he was about nineteen (1642-3). It epitomises the approach of the mechanistic tradition, and Descartes himself had its principles explained to him on a visit to Pascal in 1647. For our computerised age, it has a particular interest, since it is a form of early computer.

An experiment made at two high places, the one about 500 fathoms higher than the other

If one takes a balloon half-filled with air, shrunken and flabby, and carries it by a thread to the top of a mountain 500 fathoms high, it will expand of its own accord as it rises, until at the top it will be fully inflated as if more air had been blown into it. As it is brought down it will gradually shrink by the same degrees, until at the foot of the mountain it has resumed its former condition.

This experiment proves all that I have said of the mass of the air, with wholly convincing force; but it must be fully confirmed, since the whole of my discourse rests on this foundation. Meanwhile it remains to be pointed out only that the mass of the air weighs more or less at different times, according as it is more charged with vapour or more contracted by cold.

Let it then be set down; (1) that the mass of air has weight; (2) that its weight is limited; (3) that it is heavier at some times than at others; (4) that its weight is greater in some places than in others, as in [highlands and] lowlands; (5) that by its weight it presses all the bodies it surrounds, the more strongly when its weight is greater.[53]

Pascal conducted a series of experiments which went much further than this and were designed to expose the unreality of Aristotelian teaching on the vacuum. He attacked the accepted theory of the working of pumps, which rested on the assumption that water rose because nature abhorred a vacuum. If this was the case, he argued, why did suction pumps not raise water as high on a mountain top as at sea level, and why were two polished bodies in close contact easier to separate on a steeple than at street level.

In September 1648 one of the best-known experiments in the history of science was conducted on Puy de Dôme, near Clermont. It was recorded as follows by Pascal's relative, Perier:

The weather on Saturday last, the nineteenth of this month, was very unsettled. At about five o'clock in the morning, however, it seemed sufficiently clear; and since the summit of the Puy de Dôme was then visible, I decided to go there to make the attempt. To that end I notified several people of standing in this town of Clermont, who had asked me to let them know when I would make the ascent. Of this company some were clerics, others laymen. Among the clerics was the Very Reverend Father Bannier, one of the Minim Fathers of this city, who has on several occasions been 'Corrector' (that is, Father Superior), and Monsieur Mosnier, Canon of the Cathedral Church of this city; among the laymen were Messieurs La Ville and Begon, councillors to the Court of Aids, and Monsieur La Porte, a doctor of medicine, practising here. All these men are very able, not only in the practice of their professions, but also in every field of intellectual interest. It was a delight to have them with me in this fine work.

On that day, therefore, at eight o'clock in the morning, we started off together for the garden of the Minim Fathers, which is almost the lowest spot in the town, and there began the experiment in this manner.

First, I poured into a vessel six pounds of quicksilver which I had rectified during the three days preceding; and having taken glass tubes of the same size, each four feet long and hermetically sealed at one end but open at the other, I placed them in the same vessel and carried out with each of them the usual vacuum experiment. Then, having set them up side by side without lifting them out of the vessel, I found that the quicksilver left in each of them stood at the same level, which was twenty-six inches and three and a half lines above the surface of the quicksilver in the vessel. I repeated this experiment twice at this same spot, in the same tubes, with the same quicksilver, and in the same vessel; and found in each case that the quicksilver in the two tubes stood at the same horizontal level, and at the same height as in the first trial.

That done, I fixed one of the tubes permanently in its vessel for continuous experiment. I marked on the glass the height of the quicksilver, and leaving that tube where it stood, I requested Father Chastin, one of the brothers of the house, a man as pious as he is capable, and one who reasons very well upon these matters, to be so good as to observe from time to time all day any changes that might occur. With the other tube and a portion of the same quicksilver, I then proceeded with all these gentlemen to the top of the Puy de Dôme, some 500 fathoms above the Convent. There, after I had made the same experiments in the same way that I had made them at the Minims, we found that there remained in the tube a height of only twenty-three inches and two lines of quicksilver; whereas in the same tube, at the Minims we had found a height of twenty-six inches and three and a half lines. Thus between the heights of the quicksilver in the two experiments there proved to be a difference of three inches one line and a half. We were so carried away with wonder and delight, and our surprise was so great that we wished, for our own satisfaction, to repeat the experiment. So I carried it out with the greatest care five times more at different points on the summit of the mountain, once in the shelter of the little chapel that stands there, once in the open, once shielded from the wind, once in the wind, once in fine weather, once in the rain and fog which visited us occasionally. Each time I most carefully rid the tube of air; and in all these experiments we invariably found the same height of quicksilver. This was twenty-three inches and two lines, which yields the same discrepancy of three inches, one line and a half in comparison with the twenty-six inches, three lines and a half which had been found at the Minims. This satisfied us fully.[54]

It is possible to over-emphasise the technical skill involved in this experiment. In fact some critics have suggested that Perier 'cooked the results' by recording an unnatural consistency. For us, much of the interest lies in the function of this experiment as a weapon against Aristotelianism and the way in which Pascal used it to attack the assumptions of the organic tradition. It was almost inevitable that it should lead to a controversy which was in a sense the French equivalent of the Galilean episode. The difference was that Pascal's experiment did not conflict with orthodox interpretations of the Bible though his conclusions about Aristotelian science were as radical as any of Galileo's.

Pascal came under fire almost immediately from the Jesuit order, in the person of Père Noel, and the exchanges between the two men reveal in small compass how two different sets of assumptions may lead to differing interpretations of the same evidence. The question at issue was the nature of the space which could be observed at the top of an inverted tube of mercury when the level of the mercury fell as soon as it was inverted within a bowl of mercury. The phenomenon was a simple one but it raised a host of theological and philosophical questions. What was at stake was the authority of Aristotle in a form as crucial as the heliocentric issue, though much less dramatically. Pascal explained the gap as a vacuum, after a number of experiments with large and small tubes, which showed that the level of the mercury and hence the size of the gap varied in accordance with height above sea level and, as he argued, with the pressure of the air.

The explanation which Père Noel put forward rested upon assumptions which made sense only within an Aristotelian universe. He drew a distinction between 'natural' and 'violent' action corresponding to 'natural' and 'violent' motion. Violent action took place when a body which was united naturally was divided by an act of violence. In the case under discussion, air, which was normally a united mixture of elements, was separated by the downward fall of the mercury. Noel drew an analogy between 'air', which was a mixture of earth, air, fire and water, and blood, which was a mixture of the four humours. He assumed that one aspect of nature could illuminate another by analogy. In the new violent state created by the flow of the mercury, the element of subtle air was drawn down through the small holes in the glass from the air outside, which was disturbed in its turn. Noel used the analogy here of a sponge being squeezed in water and gradually filling out again. Squeezing was analogous to violent motion, the filling out was equivalent to natural motion.

But this argument was not enough. Noel now turned to the use of authority. He quoted Aristotle in support of his argument that a vacuum is both contrary to common sense and a self-contradiction, since it is space and non-space at the same time. He next spoke of nature in an anthropomorphic manner – it was a matter of daily experience that nature abhorred a vacuum. Finally he played the religious card. God uses nature for the ornament and variety of the world.

In his second and longer letter, where the religious note was even more explicit, Noel referred to the doctrine of the Council of Trent touching the Eucharist. He also made use of the authority of Descartes, who also rejected the idea of a vacuum.

It is quite clear that Noel regarded the 'vacuum' as a test

case of crucial significance. What seemed to be at stake was more than a merely scientific question. Roman orthodoxy and Aristotelian science were so bound up together that criticism of the science inevitably raised questions of doctrine. Perhaps also, Noel thought that a hint of unorthodoxy would be enough to frighten Pascal into silence. The controversy shows us that within the world of the Counter-Reformation the key issue was the question of intellectual authority. If Pascal rejected authority on this issue, where would his restiveness stop?

It is also fair to say that Aristotle himself laid immense emphasis on refuting the possibility of a vacuum as his answer to the atomism of the pre-Socratics. Atomism to both Aristotle and the seventeenth century was not a simple scientific theory, it was a philosophical question charged with atheistic implications. As it had come down in the ideas of Democritus and Lucretius, atomism ruled out the role of purpose within the universe. To the atomist, change was due to a random movement of atoms within an infinite universe. Where Aristotle saw final causes and pupose everywhere in the world and interpreted change as movement with a final end in view (his characteristic analogy being the acorn growing into an oak) the atomist saw only chance. The atomists' universe as sketched in Lucretius's poem *De Rerum Natura* was a fortuitous concourse of atoms moving within a vacuum.

Thus the vacuum, to some seventeenth-century minds, was a concept which admitted the possibility of atomism and hence of atheism. If there was one single factor which held up the development of an atomic theory in chemistry, it was this. Even Descartes who pushed his mechanistic interpretation to the limit and eliminated final causes from his universe drew the line at atoms. The word was too highly charged emotionally and theologically. It was the seventeenth-century equivalent of Darwin's 'natural selection'.

The correspondence between Pascal and Noel on the vacuum casts a great deal of light upon the role of scientific concepts in the seventeenth century. Noel simply could not afford to allow the possibility of a vacuum. Pascal, on the other hand, was uninterested in the theological implications of his experiment. His *Pensées* show him willing to abandon all evidence of divine purpose from the universe apart from miracles. He saw experiments as a restricted source of knowledge from which only limited conclusions could be drawn, and thus characteristic of the limitations of human reason generally. Hence he rejected Descartes' cosmology on moral grounds as much as on scientific because Descartes took too overwhelming a view of the powers of human reason. In a sense, Pascal was a French Bacon, taking his stand on experiment against hypothesis.

This is perhaps the appropriate point to mention the name of another French mechanist Pierre Gassendi (1592-1655). Gassendi was a member of the Mersenne circle, a priest and a mathematician. His interest for contemporaries lay in his claim to have constructed a theory of atomism which could be reconciled with Christianity. He could claim to have baptised Democritus and Lucretius as Aquinas had baptised Aristotle, and his views attracted attention precisely because of the Torricellian experiment and current debate about the possibility of a vacuum. Gassendi's atomism was the only philosophical position which could explain the existence of a vacuum. Aristotelians and Cartesians were at least agreed on the fact that nature abhorred a vacuum.

Gassendi's historical significance has yet to be adequately discussed by historians, though he has been receiving a good deal of attention in recent years. To his contemporaries he loomed much larger than he did to later generations. For example, his ideas were discussed at length in England during the 1650s, among others by Walter Charleton (1619-1707), who published *Physiologia – Epicuro – Gassendo – Charltoniana* in 1654. From our point of view, however, his significance lies not so much in his specific views or in his differences from Descartes but in their common mechanism. Gassendi takes his place among the exponents of mechanistic ideas within the Mersenne circle.

Robert Boyle and English mechanism

The mechanist tradition took root in England thanks in large measure to the work of Robert Boyle (1627-91) and some of his fellow members of the Royal Society, notably Robert Hooke and Henry Oldenburg (see page 178). Boyle, youngest son of the Earl of Cork, took a great interest in experimental philosophy as a young man and during his long life he published an immense mass of material. Thanks to his monied background he was able to spend freely on expensive pieces of equipment and to act as a patron generally.

Boyle's ambition was to discover the workings of the mechanical philosophy in the world of chemistry. He believed that what appear to be qualitative differences between substances, such as colour, heat or texture, were in fact due to the mechanical action of particles. As Oldenburg put it in a letter to Spinoza in April 1663, Boyle wished to explain

that the common doctrine of substantial forms and qualities accepted in the schools rests on a weak foundation, and that what they call the specific differences of things can be reduced to size, motion, rest and position of their parts.[55]

Boyle's mechanistic outlook was linked up directly with the Torricellian experiment and with the mechanism of Mersenne. Torricelli had performed his experiment in 1644, it was repeated in France in more elaborate forms from 1646-7 onwards and it reached England at least as early as 1648. It was performed regularly in Oxford when Boyle was present and a number of related experiments were conducted in 1653 by Dr Henry Power who repeated Pascal's famous Puy-de-Dôme observations on the hills outside Halifax. The culmination of two decades of scientific activity was reached in 1661 when Boyle proved that air resisted compression in proportion to its density, i.e. the more confined the air, the stronger was its 'spring'. (In modern terminology Boyle's Law states that pressure x volume = a constant.)

From our point of view, the interest of Boyle's experiments lies in their place within the mechanistic tradition. Boyle believed that the explanation for the 'spring of the air' lay in the greater concentration of particles which occurred when the air was under pressure. His observations showed that air was behaving according to mechanical principles, and he devoted the rest of his life to seeking a similar mechanical explanation for chemical change. Leibniz in his controversy with Clarke (see page 222) in the eighteenth century looked back to Boyle as the sponsor of English mechanism. He accused Clarke, and by implication Newton, of accepting the existence of occult forces which Boyle had attempted to destroy:

Mr Boyle made it his chief business to inculcate, that everything was done mechanically in natural philosophy. But it is men's misfortune to grow at last out of conceit with reason itself . . . Chimeras begin to appear again . . .[56]

The gentleman-scientist or 'virtuoso' was a characteristic phenomenon of the seventeenth century. The Honourable Robert Boyle was youngest son of the Earl of Cork as well as being 'father of English chemistry'.

THE
HONOURABLE
ROBERT BOYLE,
Esq.

F. Kerseboom p.   From an Original in the possession of D.ʳ Meade .   G. Vertue Sculp. 1738.

Boyle's influence was crucial in the spread of mechanistic philosophy in England. Without his contribution mechanism would have been associated either with the Catholics, Descartes and Gassendi, or the freethinker, Hobbes.

Part of Boyle's contribution to the mechanist cause lay in his patience and ingenuity in devising fresh experiments. In his early years (c. 1660) he made considerable use of the newly invented air pump which made a 'vacuum' possible without resorting to a tube of mercury. By means of the air pump, Boyle was able to carry out a whole range of experiments aimed at answering particular questions, for example, the effect of a vacuum on the transmission of sound. In 1669 he published the following account of an experiment designed to clear up this point, on which Mersenne had claimed to show that a vacuum did not affect the sound of bells suspended in it. Boyle showed that Mersenne was wrong:

The event of our trial was, that, when the receiver was well emptied, it sometimes seemed doubtful, especially to some of the by-standers, whether any sound were produced or no; but to me, for the most part, it seemed, that after much attention I heard a sound, that I could but just hear; and yet, which is odd, methought it had somewhat of the nature of shrillness in it, but seemed (which is not strange) to come from a good way off . . . To discover what interest the presence or the absence of the air might have in the loudness or lowness of the sound, I caused the air to be let into the receiver, not all at once, but at several times, with competent intervals between them; by which expedient it was easy to observe, when a little air was let in, the stroke of the hammer upon the bell (that before could now and then not be heard, and for the most part be but very scarcely heard) began to be easily heard; and when a little more air was let in, the sound grew more and more audible, and so increased, until the receiver was again replenished with air.

And whereas in the already published physico-mechanical experiments I acquainted your lordship with what I observed about the sound of an ordinary watch in the exhausted receiver, I shall now add, that that experiment was repeated not long since, with the addition of suspending in the receiver a watch with a good alarum, which was purposely so set, that it might, before it should begin to ring, give us time to cement on the receiver very carefully, exhaust it very deligently, and settle ourselves in a silent and attentive posture. And to make this experiment in some respect more accurate than the others we made of sounds, we secured ourselves against any leaking at the top, by imploying a receiver that was made all of one piece of glass (and consequently had no cover cemented on to it) being furnished only within (when it was first blown) with a glass-knob or button, to which a string might be tied. And because it might be suspected, that if the watch were suspended by its own silver chain, the tremulous motion of its sounding bell might be propagated by that metalline chain to the upper part of the glass, to obviate this as well as we could, we hung the watch, not by its chain, but a very slender thread, whose upper end was fastened to the newly mentioned glass-button.

These things being done, and the air being carefully pumped out, we silently expected the time, when the alarum should begin to ring, which it was easy to know by the help of our other watches; but not hearing any noise so soon as we expected. I desired an ingenious gentleman to hold his ear just over the button at which the watch was suspended, and to hold it also very near to the receiver; upon which he told us, that he could perceive, and but just perceive something of sound that seemed to come from far; though neither we that listened very attentively near other parts of the receiver, nor he, if his ears were no more advantaged in point of position than ours, were satisfied that we heard the watch at all. Wherefore ordering some air to be let in, we did, by the help of attention, begin to hear the alarum, whose sound was odd enough, and, by returning the stop-cock to keep any more air from getting in, we kept the sound thus low for a pretty while, after which a little more air, that was permitted to enter, made it become more audible; and when the air was yet more freely admitted, the by-standers could plainly hear the noise of the yet continuing alarum at a considerable distance from the receiver . . .

From what has hitherto been related, we may learn what is to be thought of what is delivered by the learned *Mersennus*, in that book of his Harmonicks (i.e. that sounds are transmitted in a vacuum).[57]

In this and many other experiments Boyle showed himself a mechanist. He also displayed a mechanistic emphasis in his religious views. Boyle believed that God's nature was revealed in his creation and hence that God was in some sense an engineer, albeit a perfect one. Boyle's God unlike Pascal's was not a hidden God but a being who delighted to reveal his power, goodness and wisdom in nature. Boyle thus made mechanism acceptable to many of his fellow countrymen in the years following 1660. In doing so, he formed part of a reaction against the religious emotionalism which had been so marked a feature of the Cromwellian regime.

In *Some Considerations of the Usefulness of Experimental Natural Philosophy* (1663), Boyle wrote of the world of nature as 'a matchless engine', revealed in such a phenomenon as the circulation of the blood, which was 'contrived' by God's wisdom.

In another treatise, *The Christian Virtuoso* (1690), Boyle spoke openly of the deity as an artificer:

We may confidently say that the Experimental Philosophy has a great advantage of the Scholastic. For in the Peripatetic schools, where things are wont to be ascribed to certain substantial forms and real qualities (the former of which are acknowledged to be very abstruse and mysterious things, and the latter are many of them confessedly occult), the accounts of nature's works may be easily given in a few words that are general enough to be applicable to almost all occasions. But these uninstructive terms do neither oblige nor conduct a man to deeper searches into the structure of things, nor their manner of being produced, and of operating upon one another; and consequently are very insufficient to disclose the exquisite wisdom, which the omniscient Maker has expressed in the perculiar fabrics of bodies, and the skilfully regulated motions of them or of their constituent parts. From the discernment of which things there is produced in the mind of an intelligent contemplator a strong conviction of the being of a divine Opificer, and a just acknowledgment of His admirable wisdom. To be told that an eye is the organ of sight, and that this is performed by that faculty of the mind which from its function is called visive, will give a man but a sorry account of the instruments and manner of vision itself, or of the knowledge of that Opificer, Who, as the Scripture speaks, *formed the eye*. And he that can take up with this easy theory of vision will not think it necessary to take the pains to dissect the eyes of animals, nor study the books of mathematicians, to understand vision; and accordingly will have but mean thoughts of the contrivance of the organ and the skill of the Artificer, in comparison of the ideas that will be suggested of both of them to him that, being profoundly skilled in anatomy and optics, by their help takes asunder the several coats, humours, and muscles, of which that exquisite dioptrical instrument consists; and having separately considered the figure, size, consistence, texture, diapheneity, or opacity, situation, and connexions of each of them, and their coaptation in the whole eye, shall discover, by the help of the laws of Optics, how admirably this little organ is fitted to receive the incident beams of light, and dispose them in the best manner possible for completing the lively representation of the almost infinitely various objects of sight.[58]

Boyle differed from mechanists like Galileo and Descartes in his lack of mathematical expertise. His contribution to the Scientific Revolution lay in devising experiments which were a series of hammer blows undermining Aristotelianism, and mechanist doctrines based on deduction rather than observation. Boyle has been called the Father of English chemistry. It seems more appropriate to call him the Father of Experimental Method. Others before him such as Torricelli and Pascal had conducted experiments. No one before Boyle devoted his life and substance to performing them, and publishing the results. It was indeed Boyle's enthusiasm and generosity which enabled the Royal Society to survive and flourish whereas its royal patron, Charles II, provided his name and nothing more. The so-called Royal Society was really Boyle's Society.

Mechanism and the Royal Society

The spread of mechanism was not entirely due to the work of individual scientists. It was also related to the formation of dedicated groups, the first of which formed itself around Galileo. After his death in 1642, its work was carried on by Torricelli and Viviani. It took on a more formal guise as the Academia del Cimento under the patronage of Leopold of Tuscany, during the years 1657-67 after which it was once more an informal group. At Bologna, there was a Anatomical Society ('Coro Anatomico') consisting of nine members. Similar groups in Paris surrounded Mersenne and later Henri Louis de Montmor (1634-79). The English equivalent of this was the group of Oxford scientists which met at Wadham College during the 1650s and later, from 1661, formed the nucleus of the Royal Society.

This belief in mechanism and a feeling of being surrounded by Aristotelian orthodoxy gave the early years of the Royal Society a missionary character. Not all the members of the Society were enthusiasts for mechanism but the three men

Thomas Sprat's *History of the Royal Society* was a
successful piece of public relations, in which the
patronage of Charles II played an important part.
The Royal Society received no more patronage from
Charles than a general blessing, but the ascription
did it as much good as any donation.

ſtitioſe quidem pleraꝗ omnia, in lu-
xum & oſtentationem mox abijſſe a-
pud gentiles: in noſtra verò religione
earum nonnulla, vt monilia & armil-
las, ad pietatem aliquos trahere cona
tos, globulos precarios in vſum pro-
duxiſſe. Nos hic globulorum ſeriem,

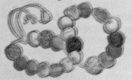

figuris eorū diuerſis exhibemus: quo-
niam in alijs aliæ effingi ſolent figurę:
rotundæ, læues, angulatæ, rhombis di
ſtinctæ. inter cæteros vnus magna ex
parte eroſus apparet: quem idcirco
addidi, quoniā talem ex gemis quas
Chalcedonios vulgò nominat, in Gal
linacei ventriculo reperi, téporis mo
ra calore eius, vt conijcio, ea parte cō
ſumptum. In duobus muſcæ apparét,
qui

Annulus cum Callimo lapillo, du-
plici facie humana inſigni, infrà pone
tur cap. II.

c. Capulis cultrorum & enchiri-
diorum aptātur Cryſtalli, Iaſpides, &
aliæ forſan gemmæ.

Cochlearia fiunt è Succino perquã
eleganti. Item ex marmore Zebii-
cio, (vt Agricola nominat ab oppido
Miſenæ iuxta quod effodit: videt aūt

Oo 2

at the centre of things. Boyle, Oldenburg and Hooke, were
very much so. They saw the Society as a missionary organisa-
tion and they published its *Transactions* as a form of missionary
pamphlet. Oldenburg's zeal in writing to kindred spirits all
over Europe was astonishing. He may have claimed that the
Royal Society had no interest in matters theological or metaphysical, but in fact mechanism was the gospel which was
being preached.

Hooke (1635-1703) was the first Curator of Experiments in
the Royal Society and he was closely associated with Boyle in
developing experiments of a mechanistic character, including
the Torricellian experiment which he performed in St Paul's
Cathedral. His associate Oldenburg mentioned this in a letter
to Boyle (25 August 1664):

Having found the top of St Paul's steeple a convenient place for
some experiments, order was given yesterday to try there the descent
of falling bodies, the Torricellian experiment and the vibrations

John Kentman of Dresden was a sixteenth-century 'geologist' whose catalogue of stones was published by Conrad Gesner in his *Rerum Fossilium* (1565). Kentman was interested in the marvels rather than the regularities of nature and he sent Gesner an account of stones found in human bodies. Gesner himself classified stones by marks on their surfaces, or by resemblances to the sun, moon or stars, or things in nature such as fruit. *Right* Kentman's cabinet in which his specimens were stored. *Left* Gesner discusses the significance of semi-precious stones. He tells his readers that such stones have been used hitherto as luxurious ornaments but now 'in our religious age' are often made into objects of piety such as the prayer beads shown here.

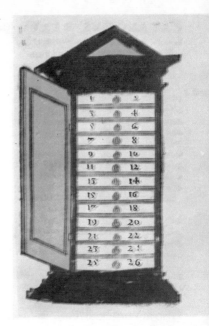

of a pendulum of the length of the top to the floor of the Church, the perpendicular height is about 200 feet.[59]

As curator, Hooke was clearly in a position to influence the course of experimental policy and it was certainly due to him, in alliance with Boyle and Oldenburg, that the Royal Society took a mechanistic line in its early years.

Hooke's mechanistic outlook appears most obviously in his book *Micrographia*, which ostensibly was a 'neutral' report of observations made with a microscope. Most of the book in fact consisted of descriptive accounts of insects and minerals, but the tone was set by Hooke's mechanistic approach. As he wrote in his description of the seeds of thyme:

... the clods and parcels of Earth are all irregular, whereas in minerals she (Nature) does begin to geometrize and practise as it were, the first principles of mechanicks, shaping them of plain, regular forms and figures, as triangles, squares, etc and tetra-haedrons, cubes, etc. But none of their forms are comparable to

Left Anthony van Leeuwenhoek's microscope (replica) *c*. 1680, 27 mm. × 47 mm. The microscope should have had an influence comparable to that of the telescope in bringing new worlds into focus, but Robert Hooke reported to the Royal Society in 1692 that microscopic studies were 'now reduced almost to a single votary, which is Mr Leeuwenhoek; besides whom, I hear of none that make any other use of that instrument, but for diversion and pastime'.

Right First pictures of bacteria seen under the microscope and described by Leeuwenhoek.

the more compounded ones of vegetables; for here she goes a step further, forming them of more complicated shapes, and adding also multitudes of curious Mechanick contrivances in their structure. (In Animals) we shall find, not only most curiously compounded shapes but most stupendious mechanisms and contrivances. (*Micrographia*, p. 154)

In describing feathers, Hooke referred to 'the mechanism of Nature' and in describing the feet of insects, he stated:

... Nature does always appropriate the instruments, so as they are the most fit and convenient to perform their offices and the most simple and plain that can possibly be ... Nor is there a less admirable and wonderful mechanism in the foot of a spider. (*Ibid.*, p. 165)

Hooke's interpretation of the world of nature pointed to the existence of a supreme mechanic who was responsible for the design of the universe:

... we shall in all things find that Nature does not only work Mechanically but by such excellent and most compendious, as well as stupendious contrivances, that it were possible for all the reason in the world to find out any contrivance to do the same thing that should have more convenient properties. And can any be so selfish

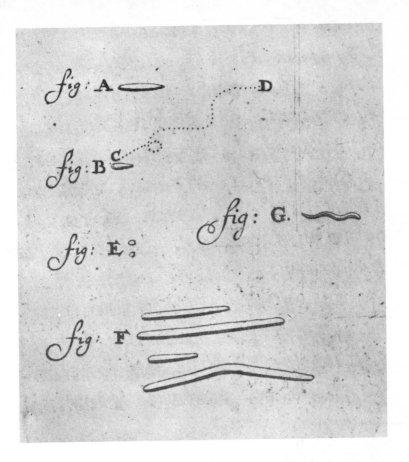

as to think all those things the production of chance? Certainly, either their Ratiocination must be extremely depraved, or they did never alternatively wonder and contemplate the works of the Almighty. (*Ibid.*, p. 171)

The interesting point here is that the details revealed by the microscope could have been interpreted by an Aristotelian as supporting acceptance of final causes in nature. Hooke, however, used them to attack the Aristotelian doctrines of matter and form, believing that progress could come only from 'the real, the mechanical, the experimental philosophy' and not from the 'philosophy of discourse and disputation', namely Aristotelianism.

By the end of the seventeenth century the mechanistic model of what an experiment ought to be had made a great deal of headway at the expense of the other two traditions. In the field of chemistry hitherto dominated by the alchemists, a new system of nomenclature was introduced by Nicolas Lemery in his textbook *Cours de Chymie* (1675). Lemery was especially critical of the use of the pretentious and obscure names given to chemical substances. Boyle in *The Sceptical Chemist* (1661) also launched an attack upon Aristotelian and alchemist alike. In these and other instances the mechanist tradition won decisive battles, though the campaign was not decided until the reforms of Lavoisier in the late eighteenth century.

Hobbes

We cannot leave English mechanism without discussing Hobbes. Thomas Hobbes (1588-1679) occupied a central position in the mechanistic group. Mersenne was a great admirer of his and Descartes sent him a copy of his *Principia* for comment. Pascal's politics recall those of Hobbes' *Leviathan*. In fact Hobbes' centrality has been underestimated by historians. His mechanistic interpretation of man and of politics was a significant attempt to push the Galilean approach into areas which had been left exclusively to the Aristotelians. Hobbes extended mechanism beyond the world of nature into ethics and pyschology. But the hostile reaction which greeted the appearance of *Leviathan* (1651) shows that the mechanistic tradition was still the affair of a small number of intellectuals.

Hobbes was a system builder in the continental tradition. He began with a mechanical version of human nature and proceeded to build a system of ethics and politics upon it. He saw movement as the essential feature of human life. Men never ceased to desire and no sooner had one wish been satisfied than another was created. The Hobbesian view of human nature was an application of the Galilean doctrine that movement, not rest, was natural. It was a world without finality and whereas the Aristotelians saw man as desiring the good, Hobbes turned this upside down in saying that men termed good that which they desired. In this universe, laws of nature were not reflections of rationality within the universe. They were simply theorems which men agreed upon for their own peace and quiet. The state was not an organic unity natural to man. It was the working of artifice, a machine, dominated by the will of a great artificer, the sovereign. These were the ideas which Hobbes absorbed during his exile in Paris and put to brilliant use.

Hobbes' mechanism was poles apart from the Aristotelian world view. It was equally so from the neo-Platonists. The Hobbesian universe was a world without mystery. Such mystery as there was, derived from mistakes in definitions which would lead the enquirer to beat his brains against imaginary walls. Hobbes had no time for the fancies, as he saw them, of an animistic world. He dismissed ghosts, fairies and witches as figments of the imagination. He reduced the role of miracles to the margin of existence and dismissed the Deity from his creation. 'When a man tells me God hath spoken to him in a dream I know he dreamt that God spoke to him'. The Hobbesian God was also a hidden God. Hobbes was a Pascal without the need for conversion.

Hobbes' work is intelligible only against the scientific background of the Mersenne circle and its successor, which enjoyed the patronage of Montmor, but he did not have the success in England which he enjoyed on the continent. Though he wrote a masterpiece of English literature, his links were with the system building of the continent. The most influential English thinkers of the seventeenth century rejected the ideas of Hobbes and, for the most part, those of his fellow mechanist, Descartes. Locke with his fundamental scepticism about the progress of science was a Baconian at heart. Not until Bentham was the tradition of Hobbes to be revived.

Part of the interest of Hobbes lies in the fact that his mechanistic interpretations of human nature, of politics and of religion struck so many of his contemporaries as implausible. They could not accept his pessimism about human nature. Seth Ward, John Bramhall and Edward Hyde all took up the cudgels against him. By implication, they placed the whole mechanistic interpretation of the world in doubt. They still accepted the Aristotelian view of ethics and of politics, since this was the tradition of the gentleman, strongly embedded in the fabric of English education and of English religion.

The same was true on the continent. The strength of the aristocratic code of behaviour and of the hierarchical view of society placed an immensely strong barrier in the way of mechanism. It was not to be expected that Europe should change its mind overnight because of a dangerous and, as it seemed, dubious hypothesis. The victory of mechanism was not to take place until the very end of the seventeenth century.

6 The great amphibium:
Sir Isaac Newton

At first sight Newton (1642-1727) appears to belong without question to what we have termed the mechanist tradition. In his *Principia* (1687), he laid the foundations for a view of the universe in which the planets moved in accordance with the same laws as governed a falling body on earth. He brought the planets and the smallest pebble within the same comprehensive scheme of things, and his synthesis was to be taken over by Voltaire and the Founding Fathers of the Enlightenment as the basis for a mechanical philosophy. The Newtonian God became the detached Deity of the eighteenth century. The universe was seen as a gigantic piece of clockwork machinery. Human nature was described in terms of quasi-mechanical responses to pleasure and pain. In all of this, Newton's influence seems unquestionable.

The mechanistic aspect of Newton's mind was also displayed in his experiments on light, first performed in 1666 and published many years later under the title *Optics* (1704). Newton's experiments showed that light behaved according to mechanical laws when passed through different media. He also showed that white light was composed of rays of primary colours:

Whiteness is the usual colour of light: for light is a confused aggregate of Rays inbued with all sorts of colours. [60]

Hence he undermined the basis of a mystical view of white light as a spiritual symbol. Indeed he came to regard light as explicable only in terms of a stream of corpuscles or particles. Newton's experiments fitted very well into Boyle's plan to explain all natural phenomena by size, shape or motion.

In his early intellectual development, Newton was undoubtedly influenced by the mechanists. As a mathematician he owed a vast amount to Descartes, whose *Geometry* and

Dioptric he read in 1664. He also studied the work of another mechanist, the *Micrographia* of Robert Hooke, and carried out some of Robert Boyle's experiments in mechanical philosophy. Most conclusive of all are the many references to Galileo in the *Principia*. Galileo's concepts of acceleration and his experiments with falling bodies had a decisive influence upon him.

In the Preface to the first edition of the *Principia* Newton claimed to have explained the motions of the planets, the comets, the moon and the sea according to mechanical principles. He went on to say:

I wish we could derive the rest of the phenomena of Nature by the same kind of reasoning from mechanical principles, for I am induced by many reasons to suspect that they may all depend upon certain forces by which the particles of bodies, by some causes hitherto unknown, are either mutually impelled toward one another, and cohere in regular figures, or are repelled and recede from one another.

Newton hoped that mechanical principles would illuminate all aspects of Nature.

Despite all this mechanism Newton drew back from a fully mechanical explanation of the Universe. He refused to draw the conclusion that 'mere mechanical causes' could give rise to the regular motions of the six primary planets and of the ten moons which revolved around the earth, Jupiter and Saturn. He felt that

This most beautiful system of sun, planets and comets could only proceed from the counsel and dominion of an intelligent and powerful Being.

This was obvious even in the disposition of the fixed stars which had been placed at immense distances from one another lest gravity should cause them to fall on each other.

A portrait of Isaac Newton (1642-1727)
as a pillar of the 'establishment'.
He did not suffer fools gladly,
and rivals not at all.

This aspect of Newton's thought obviously makes it difficult to describe him as a thoroughgoing mechanist. There seems to be an ambivalence in Newton's approach which links him with more traditional modes of thinking. Thirty years ago the economist John Maynard Keynes (who acquired Newton's manuscripts), came to the conclusion that Newton was more in the tradition of medieval alchemy than eighteenth-century mechanism.

'Why do I call him a magician?' Keynes asked:

Because he looked on the whole universe and all that is in it *as a riddle,* as a secret which could be read by applying pure thought to certain evidence, certain mystic clues which God had laid about the world to allow a sort of philosopher's treasure hunt to the esoteric brotherhood. He believed that these clues were to be found partly in the evidence of the heavens and in the constitution of elements (and that is what gives the false suggestion of his being an experimental natural philosopher), but also partly in certain papers and traditions handed down by the brethren in an unbroken chain back to the original cryptic revelation in Babylonia. He regarded the universe as a cryptogram set by the Almighty – just as he himself wrapt the discovery of the calculus in a cryptogram when he communicated with Leibniz. By pure thought, by concentration of mind, the riddle, he believed, would be revealed to the initiate.[61]

This judgment seemed to be an outlandish paradox, even though it was based upon the evidence of Newton's own notebooks. E. A. Burtt, however, in his book *Metaphysical Foundation of Modern Science* (1932) pressed home a similar point. Burtt showed that Newton's theological opinions lay behind his scientific concepts of absolute space and time. Somewhat similar arguments were put forward by Alexander Koyré in his brilliant book, *From Closed Space to Infinite Universe* (1957).

More recently still (1966) two British historians Rattansi and McGuire, in an article entitled 'Newton and the Pipes of Pan', showed that Newton proposed to include within his *Principia* a statement of his religious and historical beliefs. The passage is worth quoting because it reveals Newton as a believer in Pythagorean philosophy and the ancient theology (*prisca theologia*). In a draft scholium to Proposition VIII of the *Principia* Newton stated that Pythagoras anticipated him in discovering that the force of gravity varied inversely with the square of the distance:

For Pythagoras, as Macrobius avows, stretched the intestines of sheep or the sinews of oxen by attaching various weights and from this learned the ratio of celestial harmony ... the proportion discovered by these experiments, on the evidence of Macrobius, he applied to the heavens and consequently by comparing those weights with the weights of the Planets and the lengths of the strings with the distances of the Planets, he understood by means of the harmony of the heavens that the weights of the Planets towards the sun were reciprocally as the squares of their distances from the sun.[62]

In a scholium to Proposition IX, Newton states that the ancient attributed the perfect art and wisdom in Natural laws to the working of the Divine Power.

To some such laws the ancient philosophers seem to have alluded when they called God Harmony and signified his actuating power harmonically by the God Pan's playing upon a Pipe and attributing musick to the spheres, made the distance and motions of the heavenly bodies to be harmonical and represented the Planets by the seven strings of Apollo's Harp.[63]

The implications of these passages are crucial to our assessment of Newton. We may now place him, partly at least, in the magical tradition of science. There is no longer any need to make a sharp distinction between Newton the scientist, looking forward to the bright future of modern science, and Newton the historian and alchemist, obsessed for some strange reason with the outmoded practices of the past. Newton was in short a second Kepler, whose scientific insights derived from his beliefs about the world as a whole. The *Principia* was not a detached piece of research. It was part of a religious and historical synthesis, the work of a great system builder.

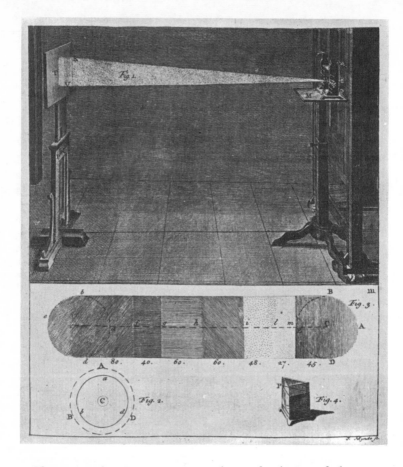

If we see the *Principia* as a piece of science of the same character as Kepler's *Harmonice Mundi*, Newton's reluctance to commit himself to a mechanistic view of the cosmos becomes intelligible. To us, all the implications of Newton's discoveries point to the conclusion that the world was a machine. But Newton insisted that this was not so. He saw God as continuously involved in the preservation of the universe, constantly correcting the slight errors which left to themselves could bring disaster. Newton's God was not a mechanic. His presence was part of the nature of things.

This illustration is from a popular work on Newtonian science, published in the eighteenth century, showing how the rainbow was formed. Part of the aftermath of Newton's discoveries was the interest in colours shown by eighteenth-century poets.

In fact Newton looked upon space and time as part of the Divine presence in the universe. Absolute space was the *ensorium* of divinity. In a famous controversy with Leibniz, Samuel Clarke defended Newton's point of view on this. Newton's cosmos was not a secular, godless creation. It was permeated through and through with the Divine Presence. Newton was in fact a neo-Platonist.

In the fifty odd years which elapsed between Kepler's death in 1630 and the writing of the *Principia* (1687), the intellectual climate of Europe had changed a good deal. Perhaps the most striking development was the rise of the mechanical philosophy, in which the decisive instrument had been Descartes. So far as England was concerned, the mechanistic views of Hobbes were equally if not more important. Hobbes' mechanism as expounded in *Leviathan* (1651) and other writings became the great bogy of England during the second half of the seventeenth century, and the term 'Hobbist', a piece of verbal abuse among intellectuals.

To defend the world of the spirit became the major preoccupation of many theologians and scientists, prominent among whom were the Cambridge Platonists (more properly, as we have suggested, neo-Platonists), Henry More, Benjamin Whichcote, Ralph Cudworth and several others. In most historical studies of them, the Cambridge Platonists have appeared as the apostles of toleration, light and enlightenment. But this was not the whole truth. A man like Henry More was obsessed with the notion of *theologia prisca*, the ancient theology as revealed in the Cabbala, the neo-Platonists and, with due reservations, Trismegistus. The world was a living soul not a dead machine. More, it is true, originally welcomed Cartesianism with enthusiasm, but as the implications of Descartes' views became clear, More drew back. Both he and

An illustration showing one of the most famous experiments in the history of science: Newton's prism experiment in which he showed how sunlig is split into its separate colours. The result, among other things, was to take away the mystic significance which was attached to light.

Ralph Cudworth were the main English defenders of a neo Platonic point of view in the second half of the seventeent century. And their chief target was Hobbes.

Newton was educated at Cambridge in the 1650s and h read the Cambridge Platonists. His notebooks reveal th kind of interest in the past which neo-Platonists encourage He was interested in the Cabbala for example, and, as w have seen, he mentions Trismegistus. We are now justifie in looking upon the *Principia* as part of this general patter in Newton's intellectual development. Indeed perhaps w should speak of Newton himself as a Cambridge Platonist.

On his scientific side, if such a distinction is possible Newton built upon the work of Kepler. In other words, was the mathematical language of the universe which attracte his attention. Newton turned to the half-forgotten, out o date, mystical scientist of the court of Rudolph II. He too over Kepler's three laws and, using Galileo's law of fallin bodies as his measure, he described mathematically the la of gravity in its application to the whole cosmos. It was statement which Kepler would have approved of – the mutua attraction of two masses varies inversely with the square o the distance between them.

To us this now appears as a piece of amazing insight. Newto had created a synthesis which swept his predecessors int oblivion. But the *Principia* did not receive the welcome whic we might expect. The Aristotelian reaction was predictabl hostile, and equally so was the judgment of the Cartesians The Cartesians, now a power to be reckoned with in Hollan and in France, dismissed Newton's thesis on the ground tha it rested upon the assumption of 'action at a distance', i short, occult forces. Thirty years after the publication of th *Principia* Leibniz attacked it, saying 'that what has happene

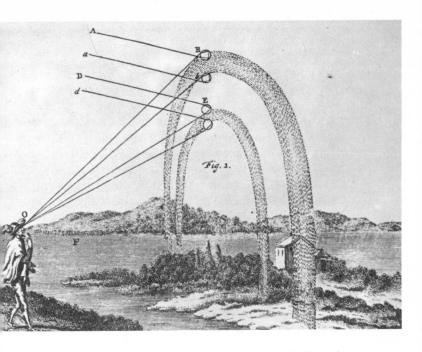

Fig. 1.

n poetry, happens also in the philosophical world. People are grown weary of rational romances . . . and they are become fond again of the tales of fairies!' Newton had taken over Kepler's ideas of the sun's power over the planets and translated into something even more fanciful the mutual attraction of masses. This was unacceptable to the new mechanists, full of a crusading Cartesianism and now on the verge of success after decades of oppression. The Cartesians rejected Newton for the same reason that Galileo and Descartes rejected Kepler. It was not a clash of 'ancient' and 'modern', as of two conflicting paradigms.

Christian Huygens, the Dutch Cartesian, dismissed Newton's principle of attraction as 'absurd' and in no way 'explicable by any principle of mechanics'. Leibniz wrote to Huygens in 1693 referring to Newton along with Aristotle as a believer in 'sympathies' and 'antipathies', which were completely implausible. Fontenelle, whose *Entretiens* became a layman's introduction to the heliocentric system, took a similar stand against Newton. It was not until Voltaire published his introduction to Newtonian ideas in the 1730s that any headway was made in France. The reason for this rejection is clear enough. Newton's theories had an aura of neo-Platonism which made them appear completely wrong-headed to many mechanists. Indeed it must be said that Newton was in some ways old-fashioned by continental standards. Historians of science seem to be generally agreed that Leibniz was much the more able mathematician and that English mathematics took a long time to recover from Newton's dominance over English science in later years, when he appointed only those who agreed with him to key positions. Those who had the misfortune to clash with him, like the astronomer Flamsteed, committed professional suicide.[64]

From our point of view the significance of Newton lies in his bringing together the mechanistic and magical traditions. In one of them, the world was a work of art and God was an artist. In the other, the world was a machine and God was an engineer. The two world pictures were clearly incompatible but Newton himself managed to meet the difficulty by creating a Deity who combined engineering skill with artistic solicitude. Newton's God was an aesthetic mechanic who was forever tinkering with his creation. This compromise barely survived Newton's death. The general trend of scientists in the eighteenth century was to see the world in terms of a machine. Newton the Great Amphibium managed to span two worlds, but his successors could not. Hence the *Principia* came to be regarded as the foundation of a mechanistic view of the universe.

7 The social background of the Scientific Revolution

So far we have kept our discussion of the three scientific traditions within the history of ideas. But we may now go further and ask the question, to what extent were these scientific traditions associated with a particular social environment? For the Aristotelian intellectual the answer seems clear enough. The Scientific Revolution, in so far as it occurred within an Aristotelian framework, was the work of Galenist physicians. The physician was a member of a small, wealthy elite, belonging to one of the three main professions of sixteenth- and seventeenth-century society. Numerically speaking, the physicians were much less numerous than clergy or lawyers, but they were alone among the three professions in being professionally involved with natural phenomena.

The rise of the physician was a social feature of the sixteenth century. He provided a medical service for aristocracy, gentry and merchants who were to be found within the cities in increasing numbers and by the end of the sixteenth century medicine had a more important place in those centres of elite education – the universities. This rise in prestige and growth in numbers did not mean that the medical profession was automatically geared towards a scientific revolution. The authority of Galen was accepted by almost all physicians. Generally, hierarchical assumptions about the universe, society and nature passed easily into medical teaching and there was no sharp division between the clerical world of Aristotelianism and the medical world of Galenism. Within this intellectual environment, the impetus towards change was relatively slight, and arose largely from the need to clarify the teaching of the accepted medical authorities, especially Galen. Medicine was affected by the Renaissance in the sense that purer texts of Galen became available and thanks to the invention of printing were more widespread. The pressure

The social difference between Vesalian and Paracelsan science is evident from a comparison of this portrait with that on page 205. Here Vesalius was self-evidently a successful member of a medical elite.

ANDREÆ VESALII.

towards change came from within the intellectual tradition.

Vesalius, who was educated at Louvain and Paris and came from a long line of physicians and court apothecaries, was in many ways typical of this highly educated medical elite. He shared contemporary enthusiasm for humanist ideas and he saw himself as following the example of Galen, whom he regarded as a man always willing to correct himself in the light of further experience. Vesalius accepted the Galenic theory of medicine though he had reservations about its teaching on the human anatomy. The advances associated with Vesalius took place within a traditional framework, which he saw himself as trying to free from later accretions.

Another typical figure was William Harvey, physician to Charles I and member of the Royal College of Physicians, a socially restrictive body. Like Vesalius, Harvey was working intellectually and socially among higher social groups. The wide social gap between the physician on the one hand and the surgeon and the apothecary on the other was accepted as the medical equivalent of social distinctions in society at large.

Thus we may conclude that the social roots of the scientific change within the Aristotelian-Galenist tradition lay in the medical profession. It was in this relatively small group that one side of the Scientific Revolution got under way. Its conservatism should not be underestimated. Harvey's discovery of the circulation of blood took a long time to find acceptance. But when all due reservations have been made, the physician of the sixteenth and seventeenth century had a role to play in the Scientific Revolution.

As we have seen, the cast of mind created within this tradition was more biological than mathematical. Harvey's outlook may be seen as characteristic in that his interest in the heart derived from the Galenic assumption that this was the

Anatomical demonstration at Leyden,
by Peter Paaw (1564-1617).
There is a moral tone about the
scene suggested by the skeleton
dominating the room and the wide
range of human types depicted.

supreme organ of the body. On the other hand, the Galenist was cut off from making other kinds of discovery. Galenists dismissed chemical cures as useless and hence played no part in chemical advances of the sixteenth and seventeenth centuries. An appropriate analogy here perhaps is with the division of modern psychology into various schools of thought, each with its appropriate insights and techniques.

With the Paracelsans, we shift our attention to a very different social environment. As suggested above (see page 114) Paracelsus, the illegitimate child of a nobleman, was an alienated scholar, who worked among the lower social groups of south Germany. As Dr Pagel puts it:

What personalities could be more contrasted than Vesalius with his curled beard, his courtly bearing, his Ciceronian eloquence, and Paracelsus, coarse and strident, with the appearance, in stature and raiment, of a barber surgeon?

Paracelsus rejected the hierarchical universe of the Aristotelians and substituted for it a system based upon the three principles of sulphur, mercury and salt. There were social implications in this as much as medical ones, and it followed as a matter of course that Paracelsan ·teaching made little headway among the medical elite of sixteenth-century Europe. Indeed Paracelsus openly attacked the pretensions of university-based medicine. Error was rooted in academic sources, he declared, 'Ye are created by the universities, by Leipzig, Tübingen, Vienna and Ingoldstadt'.

But he himself was an intellectual and not an ignorant artisan. He showed considerable knowledge of orthodox medical doctrine and often made use of it. There is good reason to believe that he moved in university circles as a young man, particularly in Italy. He claimed to be a doctor of Ferrara

TAB. VI. ✿ FELIS, ET LEPORIS. 29

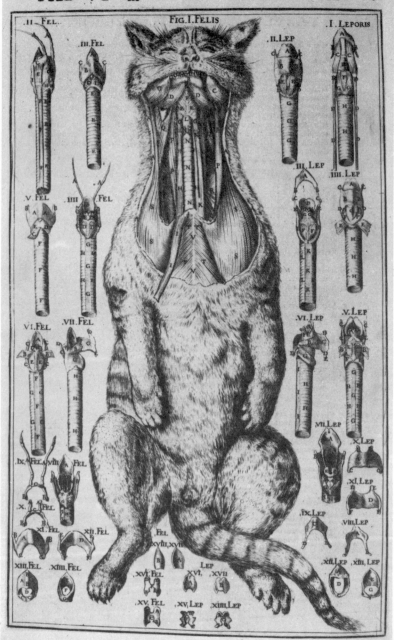

Giulio Casserio (1552-1616) *De vocis auditusque organis historia anatomica* (1601). Casserio was contemporary of Fabricius of Acquapendente at Padua. He is said to have taken the latter's place when he was sick in 1595 and 1604, but later to have suffered because of Acquapendente's jealousy. This illustration bears witness to the flourishing anatomy school at Padua *c.* 1600.

University and many of his ideas derived from the Florentine Academy. For example, he saw himself as the restorer of ancient knowledge ('prisca') overlaid by the rubbish of later doctrines, especially those of Aristotle, Galen and the Arabs.

There was also a note of religious protest in the writings of Paracelsus. The object which he had in mind was not merely a revival of medical truth but of religion as well. The social stratum for which Paracelsus spoke was that of the despised surgeons and apothecaries whose protest was expressed in religious as well as medical terms. They believed that logic had led medicine into a blind alley and also was responsible for the ills which afflicted the Christian Church. Thus Paracelsan medical teaching made headway during the second half of the sixteenth century among the surgeons and the apothecaries who were more numerous than the physicians. But the note of social and religious protest which the authorities associated, rightly, with Paracelsan doctrines led to censorship and discouragement. Few Paracelsan books were published and those which were, were signed with initials, not christian names.

The tone of alienation which we have noted in Paracelsus himself may be seen also in those who openly espoused his teaching, Bruno, for example. The same feature of restlessness may be seen in the career of another Paracelsan, John Dee, the Elizabethan astrologer. Again, Walter Raleigh, who was increasingly at odds with his political environment, was a follower of Paracelsus, so also was 'the wizard Earl' Northumberland, whose alienation took the form of Catholic recusancy. Tommaso Campanella, Bruno's fellow Neapolitan, was another alienated figure, and so also was Helmont who suffered from the attentions of the Inquisition.

This characteristic note of alienation helps to explain, at

least in part, the Paracelsans' obsession with power and with magic. We may be tempted to see them seeking in the world of nature the power which was denied to them in their daily lives. There is a complete contrast here between the restlessness of the Paracelsans and the professional assurance, not to say complacency, of the Galenists. The Paracelsans felt themselves oppressed and this led to secrecy and paranoia. The Galenists dominated the academic world of publishing and text books.

Though they shared certain aspects of the magical tradition, there was a world of difference, socially speaking, between mathematicians and Paracelsans. No note of social protest was sounded among neo-Platonists like Copernicus, Kepler and Newton. We may, if we wish, regard them as alienated from their society, Copernicus in East Prussia, Kepler the Lutheran in the world of Hapsburg patronage, Newton the heretical mathematician in orthodox Cambridge, but this alienation did not take a social form. The neo-Platonist, unlike the Paracelsan, was to be found working within the accepted institutions of society and accepting its patronage.

Where the Paracelsans turned to the pursuit of iatrochemistry, with its glorification of manual expertise, the neo-Platonist looked to the planets. The Paracelsan stressed chemical experiment; the neo-Platonists believed that the key to universal knowledge was mathematics. Hence the achievement of the neo-Platonists took place in the world of astronomy. They were not looking for power, but for understanding.

The history of neo-Platonism in sixteenth- and seventeenth-century Europe has yet to be written, though certain landmarks are clear enough, beginning with the establishment of the Florentine Academy. Ficino and those who followed his lead sought the realities of existence in the unchanging Platonic

Compare this portrait of Paracelsus,
a spokesman for lower social groups,
with that of Vesalius on page 198.

ALTERIVS NON SIT, QVI SVVS ESSE POTEST.

LAVS DEO, PAX VIVIS, REQVIES AETERNA SEPVLTIS.

OMNE DONVM PERFECTVM A DEO, IMPERF. A DIABO.

AVREOLVS PHILIPPVS THEOPHRASTVS.

AV. PH. TH. PARACELSI, NATI ANNO 1493. MORTVI ANNO 1541. AETA-
TIS SVAE 47. EFFIGIES.

Sir Jonas Moore published a *New System of Mathematicks* in 1681. The illustration reveals Moore's practical interests, as surveyor-general of the ordnance. Earlier, in 1649, he had been surveyor of the Fen drainage system. Much was expected of practical benefit from science in the early years of the Royal Society, not always with justification.

A New
SYSTEM of MATHEMATICKS
Compofed by the Eminently Worthy
Sʳ JONAS MOORE Knight
Late Surveyor Genˡ of his Maᵗˢ
Ordnance and Fellow of the
Royall Society &ᶜ.

universe, and not in the changing world of Aristotle. Burckhardt called this development 'the fairest flower of Renaissance scholarship'. From another point of view, however, it represented a retreat from the dangers and indignities of practical politics. In their aloofness, the neo-Platonists followed a different scale of values from those implied in humanist emphasis upon rhetoric and politics. Neo-Platonists rejected activity and turned to the contemplation of eternal truths, an attitude which when combined with a mathematical emphasis, is recognisably one of the roots of modern science. From this point of view, neo-Platonism and science represent the rejection by an intellectual minority of the values held by the elites of western Europe. It was an attitude, to be found in various guises from the eighteenth to the twentieth century, by which science represented a more attractive intellectual alternative to the compromises of the everyday world.

The social background of the mechanists is not susceptible to any simple explanation, though some historians have connected the rise of mechanism with the rise of a self-regulating commercial economy, such as appeared in England and Holland during the seventeenth century. This is certainly a tempting hypothesis but the evidence indicates that mechanism was propagated among circles of intellectuals with a high proportion of gentlemen among their members and few merchants. Their picture of the scientist was that of the 'virtuoso' or cultured gentleman. As Sprat put it in his *History of the Royal Society* (1667), the gentleman was free to range over learning at large without being confined to a particular specialism:

There is also some privilege to be allow'd to the generosity of their spirits, which have not bin subdu'd and clogged by any constant toyl . . . Invention is an Heroic thing and plac'd above the reach of a low and vulgar Genius.

The social aspect of mechanism is also indicated by the patronage on which the various societies relied. The Academia del Cimento was based on a Tuscan Court, the Academie des Sciences on the support of Louis XIV and the Royal Society upon the benevolence of Charles II. Typical members of these groups were gentlemen like Boyle or Towneley, physicians like Borelli, or clergymen like Mersenne. In Holland, Cartesanism spread among physicians and theologians.

What is significant is that the mechanist groups were in a state of tension with their societies, rather than reflecting the views of the great majority. Even in England, the political and social pressures towards conformity were intense. The mechanists were in a minority and remained so, though looking back we may make the judgment that they included most of the gifted men of the age.

Geographically speaking, we may see the University of Padua as the central growing point of the Galenic tradition. Throughout the seventeenth century its prestige remained high and only began to decline in the eighteenth. Paracelsan doctrines had their origins in southern Germany and Bohemia and never seem to have lost this link, even after the Habsburg victories in the 1620s. Pietism with its undogmatic emotional approach to religion, and Paracelsanism seem to have been closely associated. Neo-Platonism is more difficult to pin down but we must include the English universities and especially Cambridge, as one important centre. France also seems to have had its neo-Platonists. Finally, Holland and not least the University of Leyden, which grew in prestige during the century, became a centre of mechanism.

Chronologically also, the Scientific Revolution shows several shifts of emphasis. The decline of scholasticism was linked in part with the decline of the Habsburg powers, Spain and the Holy Roman Empire. Intellectual prestige should not follow the flag, but it often does, and the catastrophic failure of these two powers in the later years of the Thirty Years' War and the decades following had an immense impact upon their cultural influence. The same must also be said for the central European base of the Paracelsans and for the Venetian economic empire. The centre of social and economic gravity shifted to Holland, France and England. The victory of mechanism at the end of the century was largely a victory for English, French and Dutch ideas.

Puritanism and science

We may conclude this chapter by examining briefly the view that emergence of modern science is bound up with Puritanism and the rise of capitalism.

This view is worth looking at in a little more detail because it has a certain plausibility about it. It is an argument which sounds most persuasive in the mouths of Anglo-Saxon historians, English and American, because it rests upon a basic assumption that England led the way in religion, trade and science. So far as science is concerned, these historians lay particular emphasis upon the significance of Bacon and the founding of the Royal Society in 1662. They stress Bacon's devotion to experiment and the influence which this had upon the scientists of the Royal Society. By an imperceptible transition Baconian ideas come to be the criterion of scientific thought, and Francis Bacon appears as the main inspiration of the Scientific Revolution.

The next stage in this argument is the interpretation placed upon Bacon's outlook itself. The practical side of Bacon's interests is stressed to the exclusion of almost every other

Gresham College, London, first home of what was later the Royal Society, founded in the last years of Elizabeth's reign under the terms of the will of Sir Thomas Gresham (1519-79) of 'Gresham's Law' fame. He hoped that a reverse Gresham's Law would operate by which good learning would drive out bad.

aspect. Bacon's enthusiasm for knowledge which could be gleaned from unlearned sources, such as mechanics and tradesmen, is seen as a view of science resting upon practical utility. The Scientific Revolution is thus explained as a movement of largely practical significance, which rested ultimately upon the achievement of hard-headed practical men, for whom Bacon was part inspiration, part spokesman.

From now on the running is comparatively straightforward. This simplified practical interpretation of science is now linked up easily with the world of Puritanism and of commerce. Puritans are seen as men whose religious orientation derives from religious experience and hence by analogy, from 'experiment'. Bacon's supposed links with Puritanism are stressed and his interest in experiment is explained as part of his Puritan outlook.

Finally, this interpretation assumes a connection between Puritanism and the rise of capitalism. Economic individualism is seen as the obverse of religious individualism. The economic virtues of thrift and honesty appear as side effects of the Puritan 'calling'. It is taken for granted that most English merchants were Puritan and that the 'growth areas' of the English economy derived much of their impetus from the Puritan outlook.

The piece of evidence which seems to clinch the issue is the foundation of Gresham College in the last years of Elizabeth's reign. Its founder was the merchant Sir Thomas Gresham and its object is seen as teaching practical science to the merchants and artisans of the City of London. The apotheosis of Gresham College appears in the foundation of the Royal Society in the 1660s. The crucial link here was the group which made its appearance at Oxford during the 1650s under the patronage of John Wilkins, Cromwell's brother-in-law.

These men acted as midwives for the birth of the new science which was at once Puritan, Baconian and commercial.

The persuasiveness of this explanation rests upon the acceptance of certain sharply defined assumptions. As soon as these are brought into question, the solidity of the whole structure appears less convincing. Indeed the whole interpretation, to my mind, is a piece of special pleading which demands an extraordinary combination of intellectual gymnastics and rigid dogmatism.

The general argument of this book has been that whatever may be said about modern science, seventeenth-century science cannot be defined in any simple way and that we must take into account the existence of at least three distinct traditions about the way to interpret the world of nature. To use the word 'science' at all is an anachronism. To use it in a sense which confines it to a narrow interpretation of Bacon's ideas is wrongheaded in the extreme.

The proof of the pudding is in the eating. A simplist interpretation cannot make sense of major figures like Copernicus, Kepler and Descartes and it is compelled to emphasise the practical side of men like Gilbert and Newton against the general run of the evidence. Minor figures like Briggs and Harriot, and Bacon himself, are given a disproportionate significance. Indeed the place of the English contribution to the Scientific Revolution, great though it was, is made into a central feature.

To believe that Puritanism itself is a simple phenomenon or that merchants were largely Puritan is to make gratuitous assumptions which do not stand up under examination. The description 'Puritan' itself may be used to cover men of widely varying opinions and social status from ecclesiastics like Ussher to sectaries like Winstanley. It is a term so elastic that it may be used to include under one heading noblemen and artisans, from the Earl of Leicester to John Bunyan. This essential ambiguity is lost sight of in an interpretation which links up 'Puritanism' and 'science'. Indeed the basic assumption of the argument is that these two terms stand for recognisable and unified points of view, which are ultimately analogous.

The assumption that Puritanism and capitalism are closely linked is one based upon an inadequate foundation of empirical evidence. The experience of seventeenth-century England points to the fact that the great expansion in colonial trade took place after the Restoration. The kind of qualities which a merchant required for this type of economic activity were not those associated with thrift or a 'calling'. The typically successful merchant was the large-scale entrepreneur engaged in the re-export trade or in the slave trade or in trade with India. In all of these, there was need for a gambling mentality or 'flair'. Another way to economic success lay in establishing a link with Government patronage for which Puritanism was a disadvantage.

If we must associate Puritanism with some aspect of English economic life, it seems more appropriate to look at the clothing areas which from the mid-sixteenth to the mid-seventeenth century (at least) were in the midst of a prolonged economic depression, relieved only by the short boom of 1604-14. The most flourishing period for the English cloth industry was the fifteenth century and the first half of the sixteenth. It was then that 'the sheep were eating up the men' in More's memorable phrase, when the demand for wool was at its highest. In the second half of the sixteenth century, the economic areas which we may associate with Puritanism were areas of economic decline or depression. Indeed Puritanism itself, in so far as we find it in certain parishes of London and in the clothing towns of Yorkshire and Lancashire, may have provided a form of religious compensation for the dark days which had descended upon the people. In this sense, Puritanism was a kind of millenarianism which is associated in other societies with periods of deep distress. The Puritans were forward-looking in the sense of looking forward to some kind of godly kingdom on earth. Their religion did not provide the impetus for economic rationalism that Weber would have us believe, but a variety of compensations for economic decline. Thus we may expect Puritanism in some of its forms to be associated with magic more than medicine.

This expectation is borne out when we examine the Civil War period in England, when Paracelsan medicine was associated with a wave of religious enthusiasm and of social protest, directed against the legal and clerical as well as the medical elites. John Webster, one of the chaplains in Cromwell's army, wrote a pamphlet urging the introduction of Paracelsan doctrines into the universities. He suggested that 'the philosophy of Hermes, revived by the Paracelsian school' be introduced into the universities and he looked forward to a future in which

youth may not be idly trained up in notions, speculations and verbal disputes, but may learn to inure their hands to labour, and put their fingers to the furnaces, that the mysteries discovered by Pyrotechny and the wonders brought to light by chemistry, may be rendered familiar unto them . . . that they may not be Sophisters and Philosophers but Sophists indeed, *true Natural Magicians, that walk not in external circumference, but in the centre of nature's hidden secrets* . . . (*Academarium Examen*, 1654)

The name of Bacon was used by this movement but its background was European and its father figures were Paracelsus, Van Helmont and Comenius. As we have seen, this does not mean that the ideas expressed by the English Paracelsans had no contribution to make to the Scientific Revolution. Van Helmont had pointed the way to experiments in gases. Glauber was a leader in experimental techniques and there is good reason to believe that the English Paracelsans (or Helmontians) made headway in England until the great plague of 1666, but suffered a great setback then when many of their members were wiped out.

The main field of interest for the Paracelsans was that of iatrochemistry (the use of chemical substances for medical cures). Technical expertise was highly prized. Manual labour was not looked down upon but accepted as a matter of course. There was little emphasis on mathematics. The characteristic scientific achievement of the Paracelsans lay in the preparation of new compounds but this activity was carried out against a background of religious and philosophical ideas, which made it appear something more than a merely technical achievement. It was a blow for truth against falsehood, of godliness against evil, of the oppressed against the oppressor.

Thus if we seek an explanation for this aspect of the Scientific Revolution, we may find it as much in social and religious discontent as in intellectual dissatisfaction with a dominant paradigm of ideas. The background of the Paracelsans was the pursuit of the millennium. Comenius like his followers looked forward to a Golden Age of universal wisdom (pansophia). So did Bruno, so also did Campanella. If they were forward-looking, it was in terms of a future of religious harmony, not of a scientific, non-religious utopia.

There was also within 'the Paracelsan mind' a tendency to reject the intellectual habits associated with gentlemen, and in this the scholar-parson, physician and the squire were considered to be part of the same social vested interest. From the Paracelsan standpoint, the division made by the Aristotelians between 'liberal' and 'mechanical' arts confirmed the close links which existed between an ungodly scholasticism and an unjust society. Reform was necessary to bring the practical values of Paracelsan chemistry into a dominating position in education. Thus the Scientific Revolution of the Paracelsans was something more than scientific.

8 The impact of the Scientific Revolution

The intellectual impact of the Scientific Revolution may be summed up in brief as the displacement of the authority of the 'Ancients', Aristotle, Plato and the rest, by the authority of the 'Moderns', Descartes, Newton, and their followers. This was a remarkable transformation. From the twelfth century onwards, intellectual progress had seemed to lie in the rediscovery of the achievements of Greek and Roman civilisation. The Renaissance itself by making new texts available reinforced this attitude. It was now presumed to lie in the application of mathematical models and experimental method to every aspect of nature. Thus, for example, in the eighteenth century, Benjamin Franklin (1706-90) used Newtonian concepts to construct a new science of 'electricity'.

In broad terms, this amounted to the victory of mechanism over the organic and magical traditions. It was the mechanist conception of science, and experiment which came to dominate the eighteenth century. Hence the reservations which Newton and expressed in his *Principia* dropped out of sight. His most popular work was the *Optics*, in which the influence of mechanism was most obvious.

Of equal interest is the impact which the victory of mechanism had upon spheres of thought which had been dominated so far by non-mechanistic assumptions. The success of the mechanists in propagating their world view may be attributed to their creation of a new intellectual institution – 'the Philosophical Society'. The Philosophical Society, dedicated to experiments and research, became a characteristic development of the late seventeenth and eighteenth centuries. In contrast, the universities, citadels of the classical tradition, remained largely on the defensive, though it was possible for philosophical societies to be formed within a particular university by a minority of professors.

Benjamin Franklin (1706-90), one of the Fathers of the American Enlightenment, helped to set the practical note of the American scientific tradition.

The type of philosophical society which actively encouraged mechanistic experiments had its mid-century origins in the Mersenne and Montmor groups in Paris and in the Royal Society, though it is possible to find antecedents in the sixteenth century. These beginnings were small, and looking back we may exaggerate their significance in contemporary eyes. But the fashion caught on and by 1750 philosophical societies were a normal intellectual feature of provincial towns. Some of these, at Berlin and St Petersburg, were founded under state auspices, after the model of Colbert's Academie des Sciences. Others, more free and perhaps more fruitful, sprang up from local initiative. In the 1680s, for example, philosophical societies were established in Dublin, Oxford and New England. Similar societies or groups existed in France, Italy, Germany and Holland.

These societies published transactions of their proceedings with the deliberate object of reaching a wider public. The *Journal des Savants* and *Philosophical Transactions* (the organ of the Royal Society) both appeared in 1665 for the first time.

Oldenburg, secretary of the Royal Society, clearly intended *Philosophical Transactions* as the mechanist equivalent of a propaganda sheet and he never tired of stating that the aims of the Royal Society were to introduce the 'true and mechanical philosophy'. Pierre Bayle, another propagandist for the cause, though of the next generation, published a journal called *Republic of Letters.* Such publications account for much of the success of the mechanists in putting their case. They reached out beyond the academic text book to the wider literate public and, whereas scientific exchanges had been confined to letters between learned men in 1600, by the end of the century the learned periodical had taken over this role. But it was as much a propaganda instrument as a source of information. The periodicals were weapons of the Moderns against the Ancients.

Science and philosophy

The success of the mechanistic Moderns in challenging the authority of the Ancients also derived in large measure from the popular appeal of their spokesmen. Descartes' *Discourse in Method* had the enormous advantage of being at once an intellectual autobiography and a piece of informal philosophy which was readily intelligible to the laymen. In contrast, the views of the schoolmen were expounded in technical Latin in works of enormous bulk. Works in the occult tradition suffered from the disadvantage that they were intended for a highly select audience. It was the mechanists alone who produced books with an appeal to a wide public.

In France, Fontenelle (1657-1757) became a populariser for the second generation of Cartesians. In his book *Conversations about a Plurality of Worlds* (1685) Fontenelle made available in literary form the ideas which had been raised by Descartes

John Locke (1632-1704), with Newton was a key figure in the victory of the 'Moderns' over the 'Ancients' at the end of the seventeenth century

and other mechanists in the mid-century. He succeeded in 'putting across' for discussion a topic which would have led to his execution for heresy in 1600, namely that other worlds were a possibility. This was an issue with obvious implications for the Christian claim to be a unique religion. If other worlds did exist, had they also fallen and been redeemed? Fontenelle's work ran into many editions before the end of the century and its popularity is an indication of the way in which traditionalism, though solidly based in the French universities, was being outflanked in the French salons.

The same process also took place in England. Here the attack of the Moderns against the Ancients was led by the philosopher, John Locke (1632-1704). Locke was a philosopher in the new style and his *Essay of Human Understanding*, published in 1690, was written in clear, intelligible prose. Locke very much regretted that philosophy, which he considered to be nothing but the true knowledge of things, was thought unfit or incapable to be brought into well-bred company and polite conversation. In his *Essay*, he attempted to make the implications of the mechanist tradition available to the gentlemen of England.

Locke himself was in many respects characteristic of the new type of intellectual. He was both a gentleman and a student of physics. He had graduated at Oxford but he had reacted against the Aristotelian tradition. He was a member of the Royal Society and an enthusiast for experimental philosophy. Where he was very influential was in his role as the spokesman for the political views of the Whigs who came to power in the 'Glorious Revolution' of 1688. The Whig victory undoubtedly provided leverage for Locke's views in philosophy as well as in politics.

Locke specifically mentions the mechanistic moderns, Boyle, Sydenham, Huygens, as well as 'the incomparable Mr Newton', in his *Essay*. But even if he had not done so, his own modern position emerges clearly. He regarded mathematics, not logic, as the only analogy for reaching certainty.

In Locke we find a combination of faith in mathematics combined with a general scepticism about the possiblities of reason outside the mathematical sphere. The great exception was in ethics, where Locke considered that mathematical demonstration was possible. Since he also thought that morality was the main aim of human existence, this was an important conclusion to reach. But it also implied that the vast metaphysical systems which had been created by the scholastics were so much useless lumber. Men were better occupied in the pursuit of morality and of utility. Locke's concern was with 'profitable knowledge', not with abstract speculation.

Locke's *Thoughts on Education* (1703) carried still further his attack upon the Ancients. In this work, Locke openly criticised the traditional arts of logic and rhetoric. He thought that little good or profit came from studying logic as expounded in academic textbooks, though he always spoke well of Aristotle himself. He advised parents, 'if you would have your son reason well, let him read Bacon'. In rhetoric he did not rule out entirely the study of Cicero but he considered that well written modern works in English should be used to perfect style. He dismissed all systems of natural philosophy as pointless speculation and he urged pupils to turn to those like Boyle, who had advanced knowledge in particular fields by experiments or observations.

Locke's two essays, *Of Human Understanding* and *Of Education*, were the clearest statement of the modern position at the end of the seventeenth century, and to measure the intellectual impact of the Scientific Revolution, at least in England, is to measure the influence of Locke. Scholasticism was still a living tradition when he wrote and it still survived for some decades. But after the publication of the *Essays*, the Moderns had powerful weapons to hand. Locke's prose became a model to follow and his emphasis on utility led to a downgrading of poetry; where Milton looked upon poetry as a sublime pursuit, Locke saw it as a fanciful pastime.

On the other hand, we must not over-estimate the welcome given to Locke's ideas. His successes in some quarters were counterbalanced by criticism in others. He was accused by many of undermining the traditional basis of morality and of destroying the arguments for the immortality of the soul. His rejection of Aristotelian teaching on substance seemed to many to be incompatible with traditional Christian teaching on the Trinity. As one critic put it 'this atheistical shopkeeper [Spinoza] is the first that ever reduced atheism into a system, and Mr Locke is the second!' Indeed, the intellectual warfare which raged around Locke is yet another reminder, if one were needed, that the Scientific Revolution was not an exclusively scientific phenomenon but an intellectual revolution in the widest possible sense, involving God and Man, as well as Nature.

Science and theology

The implications of this revolution led to intense debate among the Moderns themselves, as much as between Moderns and Ancients. In the Leibniz-Clarke controversy of 1715-17, for example, the German philosopher Leibniz (1646-1716) took issue with the Newtonians over the theological and philosophical implications of the *Principia* and the *Optics*. The controversy began in 1705 but flared up again in 1715 when Leibniz complained to Caroline, Princess of Wales, that Newton's ideas were undermining the basis of natural religion. Newton allowed Samuel Clarke, a devoted admirer, to defend him in a correspondence which was published in 1717. The two men involved, Newton and Leibniz, were the most distinguished Moderns of the day. Newton, after the publication of his *Principia* in 1687 and his *Optics* in 1704 had become the object of intellectual near-idolatry on the part of many English intellectuals. Leibniz was a figure of immense eminence on the continent, without peer as a mathematician, and the range of his theological and scientific interests was extraordinarily wide.

In this controversy, the two men were in fact debating the religious implications of the mechanical philosophy. Both saw the universe as a watch. Clarke, echoing Newton, argued that it was a watch which needed continual Divine 'government and supervision'. Leibniz regarded this as a derogation from God's perfection, and argued that God must be a perfect watchmaker who foresaw everything, provided a remedy for everything and created a pre-established harmony and beauty.

Their argument was about the nature of Divine Providence in a mechanistic universe. Within the Aristotelian world of nature this had been much less of a problem since Aristotelian emphasis upon final causes implied Divine purpose. In the mechanistic universe final causes had no place and the theologians' room for manoeuvre, so to speak, became much narrower. The Leibniz-Clarke debate is illuminating because it shows a new range of problems coming to the forefront. It displays, in small compass, the impact of the Scientific Revolution upon Christian philosophers.

The attack on tradition

The controversy between Ancients and Moderns may be looked upon in general terms as one about the intellectual validity of tradition. Hitherto, western Europe had been a traditional society in the sense that its values and modes of thought were accepted because they derived from the past. The Reformation, revolutionary though it may have been, rested upon the idea of returning to the true Christian tradition. A traditional approach, looking back to the past for guidance, was built into the mental furniture of the Christian Churches. The revolutionary notion of scrapping the past and building anew was a primary effect of the Scientific Revolution.

There was an understandable reluctance to face the situation squarely. As we have seen, Newton himself, from one point of view, can be seen as a man returning to the Greeks for inspiration. Among most scientists, the Bible and Christian teaching continued to be recognised as the source of moral as well as religious truth. But the general enthusiasm for experiment and the search for a scientific method based on mathematical analogies of clear and distinct ideas acted as a solvent upon many areas of thought. The foundations of tradition began to collapse as soon as Cartesian tests of clarity and coherence were applied.

The intellectual changes of the late seventeenth century may be seen in the contrasting attitudes towards tradition of Bossuet and Spinoza. Jean Baptiste Bossuet (1627-1704), bishop of Meaux, was a typical figure of the French establishment. He belonged to a family which had long-standing connections with the royal administrative system and he himself was closely associated with Louis XIV. Baruch de Spinoza (1634-77) lived in obscurity in Holland earning a living as a lens grinder. He was born a Jew and was reared in an atmosphere of orthodox rabbinical scholarship, but a crisis of belief as a young man led him away from Jewish orthodoxy into radical dissent.

Bossuet was chosen as tutor to the Dauphin, for whom he wrote a *Discourse on Universal History*, in which he attempted to outline the history of the human race, with an understandable emphasis upon Divine Right monarchy, as exemplified by Louis XIV. Bossuet rested his case upon the Bible, which he saw as the source of all historical truth. For him the Scriptures were the oldest book in the world, in comparison with which so-called profane histories related fables or a mass of confused facts. The Bible provided us, in the clearest outline, with the main facts of the human story – the original state of bliss, the Fall, the corruption of the world, the Flood and the origin of human arts and crafts. Bossuet believed without question that Moses was the author of

Benedict Spinoza (1632-77), one of the greatest minds of 'a century of genius'. He applied the Cartesian approach of methodic doubt to the Bible.

Genesis or rather the co-author, since throughout he had been guided by the Holy Spirit.

This view of the world rested upon tradition and Bossuet could appeal to generations of learned theologians as the authority for his statements. But he was writing at a moment when the Cartesian principles of methodic doubt and of clear and distinct ideas were accepted by a growing number of intellectuals. In fact, we may regard Bossuet's *Universal History* as an attempt to refute, once and for all, by a massive show of authority, the treason of the clerks. The key figure among those at whom Bossuet aimed was Spinoza, whose *Tractatus Theologico-Politicus*, a radical piece of Biblical criticism, was published in 1670.

Spinoza's critique of the Pentateuch (the first five books of the Old Testament) was revolutionary in the best Cartesian manner. He rejected the traditional ascription of the Pentateuch to Moses and showed that it must have been written by someone else much later. He adopted an equally critical attitude to the message of the prophets, which by Cartesian standards were often obscure and contradictory. Finally, he ruled out the possibility of miracles on the grounds that 'nature follows a fixed and unmoveable order'. Miracles were clearly impossible in the mechanistic universe of Descartes, though Descartes himself had not drawn this conclusion.

Spinoza expounded in unmistakable terms the conflict which existed between the traditional world view of Bossuet and the mechanistic universe of Descartes. The orthodox reaction was one of amazement and horror. But the ideas expressed in the *Tractatus* could not be unsaid and they gained a more widespread currency after the publication of Bayle's *Dictionary* (1697). Pierre Bayle was born in France and had once been professor in the Protestant academy of Sedan. In 1681 he took refuge in Holland where he published his *Dictionary* some years later, as a counterblast to the forces of intolerance. Bayle took over much of Spinoza's critical spirit in articles which he wrote for the Dictionary, though in an article on Spinoza himself he adopted a more cautious attitude.

In Spinoza and Bayle we may see the beginnings of the Enlightenment of the eighteenth century, a movement which derived in large measure from the Scientific Revolution. The founding fathers of the Enlightenment adopted the methodic doubt of Descartes as one of their characteristic principles and extended its use into a range of social and religious assumptions, on a scale which Descartes himself had not attempted. The Enlightenment also adopted a relative optimism about the possibilities of human progress, which may be seen in Voltaire's rejection of Pascal's *Pensées*.

The influence of mechanical models

The impact of the Scientific Revolution may also be seen in the general acceptance of Newton's *Principia* as an intellectual model. Newton was seen as a man who had successfully shown how the complex phenomena of earth and sky could be described in terms of a single, mathematical system. His success inspired others to attempt a similar *coup* in their own fields of interest. Adam Smith, for example, in his classical synthesis *The Wealth of Nations* (1776) showed that the multifarious activities of the economy operated according to economic laws, which were the equivalent of Newton's scientific laws.

The influence of the Scientific Revolution led to the application of statistical methods to economics, under the revealing title 'Political Arithmetic'. Its inventor William Petty (1623-87) was a member of the Royal Society and a self-confessed admirer of Baconian ideals. The same scientific impulse may be seen in the phrase 'moral calculus' and the application of mathematical analogies to social analysis by means of the principle 'the greatest happiness of the greatest number'. Traditional organic images of society gave way increasingly to those based on machines.

On the other hand, we must not press the direct influence of the Scientific Revolution too far. The uncritical enthusiasm for change led to a reaction against it. The mechanistic interpretation of nature came to be regarded as unsatisfactory by many as an explanation of all phenomena. The result was the movement termed 'Romanticism' which was a turning towards the past and the particular, as against the future and the general law. But even Romanticism cannot be explained without some reference to the original Scientific Revolution.

The practical impact of science

So far we have confined our attention to theoretical questions. It is also appropriate to ask how far the Scientific Revolution fulfilled the hopes of some of its early publicists. Francis Bacon and even Descartes hoped that mankind would benefit from the growth of experiment. To men like Jonathan Swift these hopes remained largely unfulfilled. In Book III of *Gulliver's Travels* Swift ridiculed the pretensions of the experimenters by describing the activities of a philosopher who tried to extract sunlight from cucumbers. He also made fun of those who took mathematics as their model for action in human affairs.

Jonathan Swift was not an isolated case and examples of similar scepticism about science could be multiplied. But the defenders of the new science could point to certain areas in which the new methods had brought benefits. The change was most noteworthy in medicine. The eighteenth century was the Golden Age of the medical profession, precisely because of the changes introduced in the wake of the Scientific Revolution. The Baconian technique of careful observation and recording proved to be particularly fruitful in dealing with disease. Thomas Sydenham (1624-89), one of John Locke's heroes, adopted Baconian methods. Leyden became the intellectual capital of the new medicine, and the influence of its graduates spread throughout Europe.

Another field in which the Scientific Revolution in its mechanistic form had considerable influence was in agriculture and industry. It has been customary to regard the moving spirits in the Industrial Revolution as unlettered geniuses who lacked formal education. As long as this assumption persisted, the question of a possible link between the Scientific and Industrial Revolutions did not arise. Interest was concentrated instead upon the religious outlook of the early industrialists. T.S.Ashton, for example, in his classic account of the British iron and steel industry drew attention to the nonconformist beliefs of many of the early ironmasters and his observations, together with those of other historians, tended to confirm the hypothesis of Weber and Tawney that Puritanism was the decisive element in the formation of the capitalist outlook, and hence of the Industrial Revolution. This view is now open to question. Recent research has shown that nonconformists were very much a minority.

Historians have shown that many of the men involved in the early development of industry in Great Britain were not ignorant of science; on the contrary, they showed considerable interest in scientific questions. James Watt (1713-1819) benefitted from his acquaintance with the Glasgow university scientist, Joseph Black (1728-99), whose work on latent heat enabled Watt to make decisive improvements in contemporary steam engines. John Smeaton (1714-92) was a member of the Royal Society, to which he read a paper 'Experimental Enquiry concerning the Natural Power of Water and Wind'. Smeaton used laboratory experiments and theory to improve his machines and two of his water wheels were used at the Carron Ironworks, the first ironworks in Scotland. With Watt and Smeaton and many others we can see scientific analysis employed as the servant of industry. The Lunar Society, founded in Birmingham in the mid-eighteenth century included early industrialists among its members.

Science also influenced agricultural improvement. In the early eighteenth century societies were founded all over Europe with the object of agricultural improvement. The Royal Society in its early years sent out questionnaries to the various counties of England. Here as elsewhere the scale of change must not be exaggerated but there is no doubt that experimental methods fostered by scientists brought results.

Why then did the Industrial Revolution originate in Great Britain and not the Continent? If the Scientific Revolution was an intellectual phenomenon in several countries of western Europe one would expect similar consequences to occur in these as well as in England. The answer seems to lie in a divergence of scientific interest between England and the Continent. In England, owing to the influence of the Royal Society, the values of Baconianism predominated. From the beginning, the Royal Society had discussed methods of improving the technical standards of particular industries. On the Continent, in contrast, the Cartesian tradition grew ever stronger, and there was greater emphasis upon the satisfaction of intellectual curiosity, as opposed to practical utility.

The contrast was most marked in mathematics and in technical inventions. On the Continent there was an extraordinary efflorescence of mathematics and of the mathematical sciences of astronomy and mechanics, electricity and magnetism. The great names in these fields of study were nearly all European, the Bernoulli family, Euler, Lagrange, Monge, D'Alembert, Boscovich, Maupertuis, Coulomb, Lavoisier and many others. Few Englishmen ranked in this company.

In England, the influence of Francis Bacon directed men's interests away from mathematics and knowledge for its own sake towards matters of more practical movement. The number of inventions made by Englishmen during the eighteenth century was phenomenal – in iron smelting, cotton spinning and weaving, steam engines, roads, canals and transport. The wealth of inventiveness in the economic field contrasts with relative barrenness in pure science. Scientific societies in

England turned their attention to questions of practical significance, as if to meet Swift's gibes in *Gulliver's Travels*. Scotland, in contrast, was less Baconian and more continental in outlook.

The contrast between English and Continental development was not absolute. The Paris Academy of Sciences published accounts of methods used in the various arts and crafts (1761-81). Some European scientist put their theoretical knowledge to practical use in textile manufacture, the preparation of chemicals and the production of sugar from beetroot. But in general it seems true to say that Baconian influences prevailed in England, Cartesian influences on the continent.

The Scientific Revolution and the literary imagination

The effect of the Scientific Revolution upon language and the literary imagination was bound up with the replacement of the Aristotelian and occult traditions by the mechanistic tradition, a change which may be seen at its most dramatic in the rise of a new prose style in those countries affected by the Scientific Revolution. The virtues of the new prose were clarity and conciseness, seen to perfection in the work of Descartes and Pascal. In England, the new fashion had its classical exposition in the placid, unadventurous prose of John Locke. In Germany, on the other hand, where the mechanistic philosophy had little impact, a manageable modern prose did not begin to appear until the late eighteenth century. Leibniz wrote mostly in French or Latin, rarely in German.

By the end of the seventeenth century in France, England and Holland, the new prose had become the main model to be imitated. The change was yet another victory for the Moderns over the Ancients. Where Hooker had followed Cicero and Bacon had followed Tacitus, the prose-writers of the late seventeenth and early eighteenth centuries copied their own contemporaries. Even Swift, conservative though he was in many ways, achieved a modern prose style.

The analogy for the new prose style was mathematics and the change occurred about the middle of the century, during the decades when Descartes' *Discourse on Method* was beginning to exercise a decisive influence. What this involved may be seen in a simple comparison of Hobbes and Locke. Hobbes, though still a mechanist, used traditional language and imagery to describe the state of nature.

In such condition, there is no place for industry; because the fruit thereof is uncertain: and consequently no culture of the earth; no navigation, nor use of the commodities that may be imported by sea; no commodious building; no instruments of moving, and removing, such things as require much force; no knowledge of the face of the earth; no account of time; no arts; no letters; no society; and which is worst of all, continual fear, and danger of violent death; and the life of man, solitary, poor, nasty, brutish, and short. (*Leviathan*, Book I, chap. 13)

Locke described it in unemotional terms:

But though this be a State of Liberty, yet it is not a State of Licence, though Man in that State have an uncontrollable Liberty, to dispose of his Person or Possessions, yet he has not Liberty to destroy himself, or so much as any Creature in his Possession, but where some nobler use, than its bare Preservation calls for it. (*Second Essay of Civil Government*, chap. 2)

Hobbes constantly sought appropriate imagery with which to sway his readers. In urging the importance of clear definitions, he painted a vivid picture of those who are trapped by false assumptions:

not finding the error visible, and not mistrusting their first grounds, [they] know not which way to clear themselves, but spend time in fluttering over their books; as birds that entering by the chimney, and finding themselves enclosed in a chamber, flutter at the false light of a glass window, for want of wit to consider which way they came in.

Science and poetry

The victory of mechanistic philosophy also changed the way in which poets went about their business. In 1600 the Galenic and occult traditions provided poets with a wide range of accessible imagery to describe human emotion. The world of astronomy was available to the poet, since the lesser world of man and the greater world of the stars were held to be analogous. The storm scenes of *Lear* provide a classic example of the parallel between unrest in nature and torment within the human psyche.

The magical tradition provided a similar range of imagery for poets like Henry Vaughan (1621-95). Vaughan, translator of a treatise of Hermetic physic, looked upon the world as the expression of the Divine mind:

> The blades of grass, thy creatures feeding
> The trees, their leafs: the flowres, their seeding;
> The Dust, of which I am a part,
> The Stones, much softer than my heart,
> The drops of rain, the sighs of wind,
> The stars to which I am stark blind,
> The Dew thy herbs drink up at night
> The beams thy warm them at i' th'light,
> All that have signature or life
> I summon'd to decide this strife.[65]

In this tradition, the Hermetic vocabulary of *sympathy, influence, magnet, ray, signature* came naturally to the poet's pen. A close connection was assumed between the stars and the smallest plant. As Thomas Vaughan, Henry's brother, put it:

There is not a *Herb* here *below* but he hath a *star* in *Heaven* above and the *star* strikes him with her *Beame*, and says to him, Grow.[66]

In the new scientific world of the late seventeenth and early eighteenth centuries, these traditional sources of language and imagery were denied to the poet, unless he used them in a fanciful or satirical manner. Poetry did not die at the touch of cold philosophy, as the Romantic poets of the nineteenth century thought, but it did change its emphasis and point of attack. The characteristic poetry of the early eighteenth century was either satire like Pope's *Dunciad* or nature poetry such as Thomson's *The Seasons* or Young's *Night Thoughts*.

The Seasons is an interesting example of the way in which Newtonian optics led the poet to appreciate the value of colour in the universe. Newton, in his *Optics*, had shown that white light was in fact made up of the colours of the rainbow. Thomson and other poets were as influenced by this fact as much as Shakespeare had been by time or Milton by space. The poetry of the early eighteenth century is distinguished by an emphasis upon the various hues which go to make up a wonderfully varied nature.

One final consequence of the decline of the magical tradition may be noted, namely the disappearance of witchcraft. Professor Hugh Trevor-Roper in his brilliant essay *The Witchcraft Craze* has noted how the rise of neo-Platonism in the sixteenth century encouraged belief in the existence of witches. Thus from one point of view the magical tradition was not a harmless piece of eccentricity but formed part of an intellectual outlook which led at least indirectly, to the persecution of thousands of helpless human beings, men as well as women. In the new mechanistic universe of the late seventeenth century witches could not exist. The victory of the machine brought about the death of witchcraft.

Thus the Scientific Revolution had an extraordinary impact upon the way in which many intellectuals thought and felt about the world. These men were still in a minority in western Europe and they were themselves divided about the implications of the new science. But the tone of the eighteenth century was to be set by them, and though the Romantic Movement was a reaction against their ideas, there was to be no real return to the traditions of 1600. The traditionalists of the nineteenth century were so only in name since their tradition was a self-conscious artifact. The Scientific Revolution created a vast gulf between traditional and modern attitudes. The past with all its virtues had gone forever. Modernity, with all its drawbacks, had been created.

Notes

1 Aristotle's *Physics*, Book VIII, 9, W.D.Ross, ed. *Works of Aristotle*, vol. II, Oxford, 1930.
2 A.R.Hall and M.Boas Hall, edd. *The Correspondence of Henry Oldenburg*, iii, p. 164, Madison and London, 1966.
3 *Opera Omnia*, tr. J.Constable (1662), p. 109, quoted by J.R. Partington in *A History of Chemistry*, vol. ii, p. 223, London, 1961.
4 F.Sherwood Taylor, 'Alchemical Papers of Plot', *Ambix*, vol. IV, p. 73, Dec. 1949.
5 Galileo, *The New Sciences*, 'Third Day', tr. H.Crew and A.de Salvio, p. 178-9, 1914.
6 *Ibid*, p. 64.
7 Quoted I.B.Cohen, *Birth of New Physics*, p. 17, 1961.
8 *Ibid*, p. 18.
9 *Two New Sciences*, 'First Day', *ed. cit.*, p. 61.
10 Harvey, *Motion of the Heart* (*De Motu Cordis*), tr. R.Willis, chap. viii, London, 1848.
11 *Ibid*.
12 Quoted J.Brodrick, *Robert Bellarmine*, vol. I, p. 71, 1928.
13 Quoted A.Koestler, *The Sleepwalkers*, p. 149, London, 1959.
14 Quoted T.S.Kuhn, *The Copernican Revolution*, p. 128, Cambridge, Mass., 1951.
15 *Ibid*, p. 129.
16 *Ibid*, p. 130.
17 D.P.Walker's article 'Kepler's Celestial Music', *Journal of the Warburg and Courtauld Institutes*, 1967, is most illuminating. So too is Miss Yate's article 'The Hermetic tradition in Renaissance Science', ed. C.S.Singleton, *Art, Science and History in the Renaissance*, Baltimore, 1967.
18 Quoted A.Koestler, *The Sleepwalkers*, p. 157.
19 Quoted E.Rosen, 'Galileo's Misstatements about Copernicus', *Isis* 49, p. 324, 1958.
20 *Ibid*, p. 324.
21 *Ibid*, p. 326.
22 Quoted F.Yates, *Giordano Bruno and the Hermetic Tradition*,

p. 208, London 1964.

23 P. Fleury Mottelay, tr. Gilbert *De Magnete*, Book v, chap. 12. London, 1893.

24 *Ibid.*

25 *Ibid.*

26 Quoted F. Yates, *Bruno*, p. 188.

27 George Peele, 'Honour of the Garter', quoted R. H. Kargon, *Atomism in England*, p. 12, Oxford, 1966.

28 Quoted P. M. Rattansi, 'Alchemy in Ralegh', *Ambix*, vol. XIII, p. 127, 1965-6.

29 Quoted W. Pagel, *Paracelsus*, pp. 142-3, New York, 1958.

30 Quoted J. R. Partington, *A History of Chemistry*, vol. II, pp. 234-5.

31 Quoted J. Mepham, 'Helmont', in R. Harré, ed. *Early Seventeenth-century Scientists*, p. 141, Oxford, 1965.

32 Quoted M. Caspar, *Kepler*, p. 63, London and New York, 1959.

33 *Ibid*, p. 267

34 *Ibid*, p. 280.

35 Quoted A. Koestler, *The Sleepwalkers*, pp. 258-9.

36 Caspar, *ed. cit.*, p. 95.

37 *Ibid*, p. 267.

38 Koestler, *ed. cit.*, p. 265.

39 G. de Santillana ed. *Galileo's Dialogue on the Great World Systems*, p. 452, Chicago, 1953.

40 *Ibid*, p. 415.

41 *Ibid*, p. 3.

42 *Ibid*, p. 35.

43 *Ibid*, p. 114.

44 *Ibid*, p. 71.

45 *Ibid*, p. 94.

46 *Principia Philosophiae*, part IV, chap. xvii, tr. Everyman edn., p. 226.

47 *Ibid*, p. 171.

48 *Principia*, part II, chap. xxxvi, E. Anscombe and P. T. Geach, edd., *Descartes: Philosophical Writings*, p. 215, Edinburgh, 1954.

49 *Ed. cit., Principia*, part iv, chap. cciv.
50 *Ed. cit., Principia*, part iv, chap. cxcviii.
51 *Discourse on Method*, Everyman edn. pp. 39-40.
52 *Principia*, part i, chap. xxviii, Everyman edn. p. 176.
53 'Treatise on the weight of the air', I. H. B. Spiers ed. *The Physical Treatises of Pascal*, pp. 31-2, New York, 1937.
54 *Ibid*, pp. 103-8.
55 A. R. Hall and M. Boas Hall, *op. cit.*, vol. ii, p. 40.
56 H. G. Alexander, ed. *Leibniz-Clarke Correspondence*, p. 92, Manchester, 1956.
57 Boyle, *A Continuation of New Experiments Physico-Mechanical*, (1669), Expt. 41, repr. J. B. Conant, *Harvard Case Studies in Experimental Science*, pp. 37-8, Cambridge, Mass., 1957.
58 Printed in P. E. More and F. L. Cross, *Anglicanism*, p. 103, London, 1935.
59 Hall and Boas Hall, *op. cit.*, vol. ii, 1663-5, p. 208.
60 H. W. Turnbull, ed. *Correspondence of Isaac Newton*, vol. i, p. 98, Cambridge, 1959.
61 J. M. Keynes, *Essays in Biography*, pp. 313-4, London, 1951.
62 Quoted J. E. McGuire and P. M. Rattansi, 'Newton and the Pipes of Pan', in *Notes and Records of the Royal Society*, p. 116, Dec. 1966.
63 *Ibid*, p. 118.
64 Professor Frank Manuel has recently described the unpleasant side of Newton in *A Portrait of Isaac Newton*, Cambridge, Mass., 1968.
65 Henry Vaughan, 'Repentance', L. C. Martin ed. *Works*, p. 449, Oxford, 1957.
66 Quoted C. H. Hutchinson, *Henry Vaughan*, pp. 153-4, Oxford, 1947.

Bibliography

Several books may be recommended to the general reader as excellent introductions to the history of science during the sixteenth and seventeenth centuries. Of these, H. Butterfield, *Origins of Modern Science* (London, 1949) is the only one to be written by a general historian and has the advantage of bringing science into a wide historical perspective. G. C. Gillispie, *The Edge of Objectivity* (Princeton, 1960) is a brilliant discussion of scientific development from Galileo onwards. Two other well written accounts of the history of astronomy and of theories of matter are S. Toulmin and June Goodfield, *The Fabric of the Heavens* (London, 1961) and *The Architecture of Matter* (London, 1962), by the same authors. A. R. Hall in *The Scientific Revolution* (London, 1954) provides a substantial account in brief compass, with his main emphasis upon scientific questions proper. T. S. Kuhn, *The Structure of Scientific Revolutions* (Chicago, 1962) is a highly original book, which seeks to explain the Scientific Revolution of the sixteenth and seventeenth century in terms of a wide-ranging theory of intellectual change. Arthur Koestler, *The Sleepwalkers* (London, New York, 1959) was criticised at some length in *Isis* (1959) for his allegedly unfair attitude to Copernicus, Kepler and Galileo. Even if these criticisms are accepted, this book remains a stimulating account of the Scientific Revolution from an unusual viewpoint. Other excellent general books are those by Morris Kline, *Mathematics in Western Culture* (New York, 1953) and L. W. H. Hull, *History and Philosophy of Science* (London, 1959). See also H. F. Kearney *Origins of the Scientific Revolution* (London, 1964).

As I have tried to indicate in this book, the scientific traditions of the Greco-Roman world and of the Middle Ages were very much a part of the Renaissance world picture and of the Scientific Revolution. On Greek and Roman science, Professor S. Sambursky in *The Physical World of the Greeks* (London, New York, 1956) and *The Physical World of Late Antiquity* (London, New York, 1962) is a superb guide. G. de Santillana, *The Origins of Scientific Thought* (Chicago, 1961) is also very illuminating. Other very useful guides on the thought of the ancient world are M. Clagett, *Greek*

Science in Antiquity (New York, 1956) and W. H. Stahl. *Roman Science: Origins, Development and Influence to the Late Middle Ages* (Madison, 1962).

J. Needham, *Science and Civilization in China* (Cambridge, 1954) is a magnificent piece of intellectual history (see the review article on it by A. R. Hall, *Economic History Review*, 1968). G. H. C. Wong, 'China's Opposition to Western Science during Late Ming and Early Ch'ing' (*Isis*, 1963) is interesting.

On medieval science, a great deal of work has been done since the pioneering work of P. Duhem, *Le système du monde* (Paris, 1913-54). The best guides in English to this difficult terrain are A. C. Crombie, *Augustine to Galileo* (London, 1952, New York, 1959) and M. Clagett, *The Science of Mechanics in the Middle Ages* (Madison, Wisconsin, 1959) (see also Clagett's article 'The Impact of Archimedes on Medieval Science' *Isis*, 1959). Of recent articles on medieval science, the following are of particular importance: A. C. Crombie, 'Quantification in Medieval Physics' *Isis*, 1961 and E. Grant 'Nicole Oresme and his De proportionibus proportionum' *Isis*, 1960. The classic work of Annalese Maier, *Studien zur Naturphilosophie der Spätscholastik* is not yet available in English translation. Another substantial book on ancient and medieval science is *The Mechanization of the World Picture* by E. J. Dijksterhuis (English trans. Oxford, 1961).

On the Scientific Revolution as a whole the standard work of reference is now R. Taton, ed., *The Beginnings of Modern Science from 1450 to 1800* (English trans. New York, London, 1964) H. T. Pledge, *Science since 1500* (London, 1966) is a shorter, but valuable, work of reference. M. Boas, *The Scientific Renaissance 1450-1630* (London, 1962) is a valuable introduction to the period. W. P. D. Wightman, *Science & the Renaissance* (Edinburgh, 1962) is a more detailed treatment and of particular interest to scholars. On the seventeenth century A. R. Hall, *From Galileo to Newton 1630-1720* (London, 1963) and I. B. Cohen, *Birth of a New Physics* (New York, 1960) have a great deal to offer the general reader. J. H. Randall, 'Scientific Method in the School of Padua' in Wiener and Noland *op. cit.* p. 139, and E. A. Underwood, 'Early Teaching of Anatomy at Padua' *Annals of Science* (Mar. 1963) are important articles while D. Stimson, *The Gradual Acceptance of the Copernican Theory of the Universe* (New York, 1917) is still relevant.

There are a number of books which do not fit into any simple category but which nevertheless are of great interest. They include M. Clagett ed., *Critical Problems in the History of Science* (Madison, Wisconsin, 1959) and A. C. Crombie ed., *Scientific Change* (London, 1963), which are valuable collections of papers read at two international conferences, with notes of the accompanying discussions. J. B. Conant ed., *Harvard Case Studies in Experimental Science* (Cambridge, Mass. 1937) offers an interesting approach to the history of science by using documentary material to illuminate the solution of certain scientific problems. P. P. Wiener and A. Noland edd., *Roots of Scientific Thought* (New York, 1957) is an excellent anthology of important articles from the *Journal of History of Ideas*. There is also Lynn Thorndike, *History of Magic and Experimental Science* (New York, 1923-58) which provides an exhaustive examination of aspects of early science which are usually ignored. M. Crosland, *Historical Studies in the Language of Chemistry* (London, 1962) is an unusual and original approach to the early history of chemistry.

On what I have called 'the magical tradition', Alexander Koyré has written a fascinating book, *Mystiques, spirituels, alchemistes* (Paris, 1955). P. H. Kocher, *Science & Religion in Elizabethan England* (San Marino, Calif., 1953) is an interesting book on a similar theme. Important recent articles on this aspect of science include: C. A. Patrides, 'The numerological approach to cosmic order during the English Renaissance' *Isis*, 1958, P. M. Rattansi, 'Alchemy and Magic in Raleigh's History of the World' *Ambix*, 1966 and J. W. Shirley, 'The Scientific Experiments of Sir Walter Raleigh, the Wizard Earl & the three Magi in the Tower 1603-17' *Ambix*, 1949.

Certain scientific figures of the period have attracted more comment than others. More has been written on Copernicus, Kepler and Galileo than anyone else. The standard work on Copernicus is now T. S. Kuhn, *The Copernican Revolution* (Cambridge, Mass., 1957) and there is also a convenient paperback edition of three Copernican treatises edited by E. Rosen (New York, 1939). The standard biography of Kepler is Max Caspar, *Johannes Kepler* (English trans. London, 1959). In a recent article, 'Kepler's Laws of Planetary Motion 1609-1666' *British Journal for History of Science* (1964), J. L. Russell examines the history of Kepler's theories after 1609.

There is now a full length biographical study of Galileo in English, L. Geymonat, *Galileo Galilei* (New York, London, 1967, English trans. S. Drake). There are also (at least) two good books on his general achievement, A. Koyré, *Etudes galiléennes* (Paris, 1939) and E. McMullin ed., *Galileo: Man of Science* (New York, London, 1967) and an interesting study of Galileo's relations with the Papal curia by G. de Santillana, *The Crime of Galileo* (Chicago, London, 1955). Santillana has also edited the Salusbury translation of Galileo's *Dialogue on the Great World Systems* (Chicago, 1953). There is a convenient paperback edition of Galileo's shorter writings by Stillman Drake, *Discoveries and Opinions of Galileo* (New York, 1957). A. Koyré, *Metaphysics and Measurement* (Cambridge, Mass., 1908) contains three essays on Galileo. Of recent articles on Galileo, the following are of particular interest: E. Grant, 'Bradwardine and Galileo' *Archives for History of Exact Sciences*, vol. 2, 1962-6, W. C. Humphreys, 'Galileo, Falling Bodies and Inclined Planes' *British Journal for History of Science*, 1966-7, A. Koyré, 'Galileo and Plato' in Wiener and Noland *op. cit.* p. 147, and S. Sambursky, 'Galileo's attempts at a Cosmogony' *Isis*, 1962.

Once off this favourite historical highway, the going becomes harder. There is a shortage of good general books on such figures as Gilbert, Bacon, Harvey, Van Helmont, Torricelli, Pascal or even Descartes, though there are many excellent chapters in individual books. On Leonardo, there is a useful symposium, *Leonardo da Vinci et l'expérience scientifique au XVI siècle Paris 4-7 juillet 1952* (Paris, 1953). The standard biography of Vesalius is C. D.

O'Malley, *Andreas Vesalius of Brussels 1514-64* (Berkeley, 1964). (See also Professor O'Malley's 'A Review of Vesalian Literature' *History of Science* 1965). On Paracelsus there is a fine book by W. Pagel, *Paracelsus* (Basel, New York, 1958), and an interesting article by W. Pagel and P. Rattansi, 'Vesalius and Paracelsus' in E. A. Underwood ed., *Science, Medicine and History* (essays presented to Charles Singer, Oxford, 1953). There is no good book on Gilbert, but his *De Magnete* is now available in paperback and there is a stimulating article by E. Zilsel, 'Origins of Gilbert's Scientific Method' in Wiener and Noland *op. cit.*, p. 219.

For Descartes, there is a very helpful chapter in Mary B. Hesse, *Forces and Fields*, (London, 1961). E. S. Haldane and G. R. T. Ross edd., *Philosophical Works of Descartes* (Cambridge, 1911) is now available in paperback as is a recent translation by J. Lafleur of the *Discourse on Method, Optics, Geometry and Meteorology* (New York, 1968). The work of Mersenne is discussed at length in R. Lenoble, *Mersenne ou la naissance de mécanisme* (Paris, 1943). On Harvey, there is Sir Geoffrey Keynes's biography (Oxford, 1966), and a number of important articles, notably: D. Fleming, 'Galen on the Motions of the Blood in the Heart and Lungs' *Isis*, 1955, D. Fleming, 'William Harvey and the Pulmonary Circulation' *Isis*, 1955, W. Pagel, 'William Harvey and the Purpose of Circulation' *Isis*, 1951 and J. S. Wilkie, 'Harvey's Immediate Debt to Aristotle and to Galen' *History of Science*, 1965. Gweneth Whitteridge has edited and translated Harvey's *De Motu Locali Animalium* (Cambridge, 1959) and his *Anatomical Lectures* (London, 1964). On Bacon, P. Rossi. *Francis Bacon: From Magic to Science* is now available in English translation (London, 1968). On Boyle, there is an illuminating introduction to certain aspects of his work by M. Boas in *Robert Boyle and Seventeenth Century Chemistry* (Cambridge, 1958), and a long article by C. Webster, 'The Discovery of Boyle's Law and the Concept of the Elasticity of Air in the 17th century' *Archives for History of Exact Sciences*, 1962-6. On the relatively neglected figure of Robert Hooke, there is a good modern study by M. Espinasse *Robert Hooke* (Berkeley, London, 1956), and Mary B. Hesse, 'Hooke's Philosophical Algebra' *Isis*, (1966). Hooke's *Micrographia* is now available in paperback. On Pascal, M. A. Bera's collection, *Blaise Pascal, l'homme et l'oeuvre* (Paris, 1956) is very useful. On Torricelli, there is a valuable discussion in W. E. Knowles Middleton, *History of the Barometer* (Baltimore, 1964).

There is now a first class critical biography of Newton by Frank Manuel, *A Portrait of Newton* (Cambridge, Mass., 1968). Newton's *Principia* and *Opticks* are now both available in paperback, published by University of California Press (1962) and Dover Books (1952) respectively. Newton's *Papers and Letters on Natural Philosophy* have also been edited by I. B. Cohen (Cambridge, 1958), and his *Unpublished Scientific Papers* by A. R. and M. B. Hall (Cambridge, 1962). Professor Cohen's book *Franklin and Newton* (Philadelphia, 1956) is very illuminating on Newton, as is A. Koyré, *Newtonian Studies* (Cambridge, Mass., 1965).

Mention should also be made of the very interesting article by J. E. McGuire and P. M. Rattansi, 'Newton and the Pipes of Pan' in *Notes and Records of the Royal Society* (Dec., 1966), and of A. I. Sabra's scholarly book *Theories of Light from Descartes to Newton* (London, 1967).

The most satisfactory general discussion of the function of scientific societies during this period is still M. Ornstein, *Role of Scientific Societies in the 17th Century* (Chicago, 1928). On the early history of the Royal Society, the most convenient work is still that of Bishop Sprat, recently published in facsimile by J. I. Cope and H. W. Jones (St Louis, 1958). For future historians of the Royal Society, there is now in process of publication a definitive edition of the letters of its first secretary, Henry Oldenburg, edited by A. R. Hall and Marie Boas Hall (Madison, Wisc., 1965-   ). There are important articles by R. Schofield ('Histories of Scientific Societies') and C. Webster ('Origins of the Royal Society') in *History of Science* 1963, 1967, and a symposium on the Royal Society in *Notes and Records of the Royal Society* (1969). Other interesting articles are, W. E. Houghton, 'The History of Trades' in Wiener and Noland *op. cit.*

pp. 354-81, and F.R.Johnson, 'Gresham College: Precursor of the Royal Society' in Wiener and Noland *op. cit.* pp. 328-53. T.Hoppen is also about to publish a study of the Dublin Philosophical Society (London, Charlotteville, 1970).

On the impact of the Scientific Revolution, the nearest approach to a general survey is still P.Hazard, *La Crise de la Conscience européenne* (Paris, 1935) translated as *The European Mind* (London, Newhaven 1953). In *From Closed Space to Infinite Universe* (Baltimore, 1957), A.Koyré deals brilliantly with the implications of the new science as seen by Henry More and others. E.A.Burtt, *Metaphysical Foundations of Modern Science* (London, 1925) is a stimulating book which puts the view that modern science had quasi-religious origins. The Leibniz-Clarke controversy has been edited by H.G.Alexander, *The Leibniz-Clarke Correspondence* (Manchester, 1956). G.Buchdahl. *The Image of Newton and Locke in the Age of Reason* (London, 1961) is a useful, if brief, selection of eighteenth-century material relating to 'The Enlightenment'. Another illuminating selection was made by J.F.Lively *The Enlightenment* (London, New York, 1966).

M.Nicolson in *Newton Demands the Muse* (Prince, 1946) and *Science and Imagination* (New York, 1956) looks at the impact of science on the literary imagination. There is a solid account of English discussion on seventeenth-century science in R.F.Jones, *Ancients and Moderns* (St Louis, 1936), and much of interest too in A.Lovejoy, *The Great Chain of Being* (Cambridge, Mass., 1936) and B.Willey, *The Seventeenth Century Background* (London, 1934). R.Schofield describes an eighteenth-century scientific society in *The Lunar Society of Birmingham* (Oxford, 1963). Important articles which discuss the impact of the new science include: R.F. Lazarsfeld, 'Quantification in Sociology – Trends, Sources, and Problems' *Isis*, 1961, R.K.Merton, *Science Technology and Society in 17th Century England, Osiris* IV (1938) (reprinted 1968), and, W.T.Stearn, 'The influence of Leyden on Botany in the 17th and 18th Centuries', *British Journal for History of Science* vol. 1, (1962-3). A.E.Musson and E.Robinson's invaluable study *Science and Technology in the Industrial Revolution*, (Manchester and New York, 1969) appeared too late for me to make use of it, much as I would have liked to. T.S.Kuhn's article 'History of Science', in D.L.Sills, ed., *International encyclopedia of the social sciences*, (New York, 1968), is an excellent brief introduction to the general questions raised by this book.

Acknowledgments

I am very grateful to my colleagues, Peter Burke, John Mepham and James Shiel for their helpful suggestions on reading the draft manuscript. I should also like to thank Peter Burke, James Shiel and Richard Brown for reading the proofs.

Acknowledgment is also due to the following for the illustrations (the number refers to the page on which the illustration appears): Frontispiece, 103, 155 Giraudon; 10-11, 20, 32, 53, 65, 69, 72 (top), 73 (bottom), 99, 111, 182 Museum of History of Science, Oxford; 15 (top and bottom), 180 (left and right), 181 Bodleian Library, Oxford; 19, 33, 131, 135 Ronan Picture Library and Royal Astronomical Society; 71, 89, 109, 192, 195 Ronan Picture Library; 142, 157, 159 Ronan Picture Library and E.P. Goldschmidt and Co., Ltd.; 36, 217, 218, 225 Mansell Collection; 45, 51, 56-7, 61, 75, 120-1, 122, 123, 125, 173, 198, 201, 205 The Wellcome Trustees; 72 (bottom), 73 (top) National Maritime Museum, Greenwich; 79 Italian Institute; 102 Magnum Photos; 165 Musée du Conservatoire National des Arts et Métiers, Paris; 189 National Portrait Gallery, London.

Index

251

World University Library

Some books published or in preparation

Economics and Social Studies

The World Cities
Peter Hall, *Reading*

The Economics of Underdeveloped Countries
Jagdish Bhagwati, *MIT*

Development Planning
Jan Tinbergen, *Rotterdam*

Human Communication
J. L. Aranguren, *Madrid*

Education in the Modern World
John Vaizey, *London*

Soviet Economics
Michael Kaser, *Oxford*

Decisive Forces in World Economics
J. L. Sampedro, *Madrid*

Key Issues in Criminology
Roger Hood and Richard Sparks, *Cambridge*

Population and History
E. A. Wrigley, *Cambridge*

Woman, Society and Change
Evelyne Sullerot, *Paris*

Power and Society in Africa
Jacques Maquet, *Paris*

History

The Emergence of Greek Democracy
W. G. Forrest, *Oxford*

Muhammad and the Conquests of Islam
Francesco Gabrieli, *Rome*

The Civilisation of Charlemagne
Jacques Boussard, *Poitiers*

Humanism in the Renaissance
S. Dresden, *Leyden*

The Rise of Toleration
Henry Kamen, *Warwick*

Science and Change 1500-1700
Hugh Kearney, *Sussex*

The Left in Europe
David Caute, *London*

The Rise of the Working Class
Jürgen Kuczynski, *Berlin*

Chinese Communism
Robert North, *Stanford*

The Italian City Republics
Daniel Waley, *London*

Rome: The Story of an Empire
J. P. V. D. Balsdon, *Oxford*

Cosmology
Jean Charon

The Arts

Twentieth Century Music
H. H. Stuckenschmidt, *Berlin*

Art Nouveau
S. Tschudi Madsen, *Oslo*

Palaeolithic Cave Art
P. J. Ucko and A. Rosenfeld, *London*

Expressionism
John Willett, *London*

Language and Literature

Two Centuries of French Literature
Raymond Picard, *Paris*

Russian Writers and Society 1825-1904
Ronald Hingley, *Oxford*

Satire
Matthew Hodgart, *Sussex*

Western Languages
100-1500 A D
Philippe Wolff, *Toulouse*

Philosophy and Religion

Christian Monasticism
David Knowles, *London*

Religious Sects
Bryan Wilson, *Oxford*

The Papacy and the Modern
World
K. O. von Aretin, *Darmstadt*

Earth Sciences and Astronomy

The Structure of the Universe
E. L. Schatzman, *Paris*

Climate and Weather
H. Flohn, *Bonn*

Anatomy of the Earth
André Cailleux, *Paris*

Zoology and Botany

Mimicry in Plants and Animals
Wolfgang Wickler, *Seewiesen*

Lower Animals
Martin Wells, *Cambridge*

The World of an Insect
Rémy Chauvin, *Strasbourg*

Plant Variation and Evolution
S. M. Walters, *Cambridge*
D. Briggs, *Glasgow*

The Age of the Dinosaurs
Björn Kurtén, *Helsinki*

Psychology and Human Biology

Eye and Brain
R. L. Gregory, *Edinburgh*

Human Hormones
Raymond Greene, *London*

The Biology of Work
O. G. Edholm, *London*

Poisons
F. Bodin and C. Cheinisse, *Paris*

The Psychology of Fear
and Stress
J. A. Gray, *Oxford*

The Tasks of Childhood
Philippe Muller *Neuchâtel*

Doctor and Patient
P. Lain Entralgo, *Madrid*

Chinese Medicine
P. Huard and M. Wong, *Paris*

The Heart
Donald Longmore, *London*

Physical Science and Mathematics

The Quest for Absolute Zero
K. Mendelssohn, *Oxford*

What is Light?
A. C. S. van Heel and
C. H. F. Velzel, *Eindhoven*

Mathematics Observed
Hans Freudenthal, *Utrecht*

Quanta
J. Andrade e Silva and G. Lochak,
Paris. Introduction by Louis de Broglie

Applied Science

Words and Waves
A. H. W. Beck, *Cambridge*

The Science of Decision-making
A. Kaufmann, *Paris*

Bionics
Lucien Gérardin, *Paris*

Data Study
J. L. Jolley, *London*